Smellosophy

Smellosophy

What the Nose
Tells the Mind

A. S. BARWICH

Harvard University Press

Cambridge, Massachusetts, and London, England

First Harvard University Press paperback edition, 2022
First printing

Library of Congress Cataloging-in-Publication Data

Names: Barwich, A. S., 1985- author.
Title: Smellosophy : what the nose tells the mind / A. S. Barwich.
Description: Cambridge, Massachusetts : Harvard University Press, 2020. |
    Includes bibliographical references and index. |
Identifiers: LCCN 2020002206 | ISBN 9780674983694 (cloth) |
    ISBN 9780674278721 (pbk.)
Subjects: LCSH: Olfactory sensors. | Rhinencephalon. | Nose.
Classification: LCC QP458 .B366 2020 | DDC 612.8/25—dc23
LC record available at https://lccn.loc.gov/2020002206

For my mother
When I was little, she didn't read me fairy tales,
but Goethe

Smell
It's the only sense
for which we can create
a new stimulus
that's never been on the face of the earth
and we can perceive it.

—Linda Bartoshuk

Contents

Preface

Smell is the Cinderella of our senses. It has acquired a remarkably poor reputation throughout history. Traditionally dismissed as communicating merely subjective feelings and brutish sensations, the sense of smell never attracted critical attention in philosophy or science. The French Enlightenment philosopher Étienne Bonnot de Condillac remarked in 1754: "Of all the senses it is the one which appears to contribute least to the cognitions of the human mind."[1] And Immanuel Kant, philosophical pillar of Prussian precision, judged as follows:

> Which organic sense is the most ungrateful and also seems the most dispensable? The sense of smell. It does not pay to cultivate it or to refine it at all in order to enjoy; for there are more disgusting objects than pleasant ones (especially in crowded places), and even when we come across something fragrant, the pleasure coming from the sense of smell is fleeting and transient.[2]

It remains unclear to me what makes Kant an expert on pleasure. But in addition to philosophers, scientists also showed little regard for the human nose. Charles Darwin noted in 1874 that "the sense of smell is of extremely slight service" to humankind.[3] Olfaction lacked reputability for scholarly investment.

Things began to change in the mid-twentieth century. A growing community of researchers recognized the potential of olfaction as a new model for the study of the senses. Revolutionary developments in neuroscience fundamentally opened up possibilities regarding its methods and outlook, especially over the past few decades. Odor perception and its neural basis are becoming key to understanding the mind through the brain. It is time to adjust our somewhat stale philosophical conjectures of mind and brain to these new realities.

This book sets out to do that. It is the book I would have wanted to read when my interest in the nose—and how it communicates with the mind—sparked. There has been a tremendous amount of sophisticated research in olfaction recently. It is time to celebrate that work. *Smellosophy* is an unapologetic declaration of love to olfaction. It provides an integrated perspective on the various creative strands of research that explore what the nose knows. These strands encompass developments in neuroscience, molecular biology, genetics, chemistry, psychology, cognitive science, and philosophy, as well as expertise in perfumery and winemaking. The philosophical tenor of this book goes beyond interdisciplinary synthesis, however. It presents an outlook, not a summary, and points at new avenues by identifying the open issues in current thinking about smell. This would have been impossible without many experts lending me their time and voice to talk about the field and its development. These conversations revealed just how diverse the field is in terms of disciplines, personalities, and opinions. It was not possible to capture all the voices (sincere apologies to those not covered), and not everyone will necessarily agree with the central argument of this book.

In the end, research on the nose continues to be in flux. Discoveries are going to be made after this publication. Maybe these discoveries will shed new light on the themes in this book, either further strengthening its claims or sounding discordant notes. After all, the book must end where the future of olfaction begins. (*Here's looking at you, kid!*)

Smellosophy

Introduction
NOSEDIVE

We begin with an experiment. Hold your nose closed with your fingers, then chew a jellybean. It will taste sweet, but not much else. Now release your nostrils, swallow, and breathe out gently. Suddenly, the intense fruity note of the jellybean, ranging from strawberry to citrus, will capture your attention. This is flavor as an olfactory phenomenon.

You may be surprised to learn that humans are excellent at discriminating between several hundreds of thousands of smells, especially as different flavor components in food and drink. We do not have a good scientific explanation for how the brain makes sense of all these scents. Smells seduce or repel us, make us feel relaxed or stressed; however, we do not comprehend how the human brain creates and gives them meaning. Perhaps, for this reason, the sense of smell is hopelessly misunderstood and often misrepresented.

This book addresses popular misconceptions associated with the human sense of smell and explores new research that will likely change your views on what your nose picks up for your brain to process and your mind to perceive. By analyzing what and how scientists know about smell, we revisit a classic question in sensory perception: What does the nose tell the brain, and how does the brain understand it? While the answer remains elusive, this book offers a contemporary philosophical and historical journey into this as-yet-unresolved neuroscientific puzzle. In this exploration, *Smellosophy* identifies the uncertainties and missing knowledge in our current thinking about odor perception, including our theoretical understanding of the perceptual dimensions of smell and their links to cognition. It is the perfect arena for competing ideas about the senses as our access to reality, and it is time to consider smell as a new model of how the brain represents sensory information and for the nature of perception in general.

Smell is a highly variable sensation—variation occurs among people but also in an individual's encounter. Its versatility in qualitative experience lends olfaction an air of subjectivity. Because of this, some do not believe the nose to be a trustworthy messenger of objective reality. This view, however, now stands corrected. Your sense of smell is much more credible than you think. Recent insights into olfaction open up a new way to think about perception, including the origins of objectivity in our subjective experience. The firm opposition between subjective and objective dimensions of the senses originates in orthodox philosophical ideas that did not incorporate knowledge about the brain. Contemporary theories of mind and brain continue with a limited perspective, as they seldom look further than what meets the eye.

The persistent paradigm for the study of the senses is vision. This is not surprising. Philosophical tradition centered on understanding in terms of "seeing things." Scientists have learned more about the visual pathway than any other sense. In the late 1950s and 1960s, advances in research on the visual system resulted in a mapping approach to the brain that continues to define the field of neuroscience, as well as the growing cross-disciplinary debates about the philosophical implications of its findings. The visual paradigm centers on spatial topology and assumes that external stimuli are represented by localized neural patterns in particular functional areas. Accordingly, specialized regions of the brain create and process our perception of particular features, such as the orientation, shape, or color of an object. This organizing principle of the brain has also been found in other sensory systems, such as audition. Similarly, if we were to look into your brain and see activity in a specific part of the auditory area in the cerebral cortex, we would know whether you are experiencing a high- or low-pitched tone.

Perceptions of color in vision and sound in audition are based on one chief causal parameter (the wavelength of the electromagnetic spectrum of light in vision, and air pressure waves in audition), which can be mapped in a linear fashion onto neural correlates. The olfactory stimulus is multidimensional. Odor quality is caused by structurally diverse chemicals with several thousand molecular features. How the human brain uses neural space to encode such an array of nonspatial stimulus information is, as yet, unresolved. No map for smells has been developed for the main cortical regions of the olfactory brain to date. Whether such a map is even possible remains an open question.

Smells come in general classes, like garlicky, burnt, floral, and green; their number within a category is almost limitless. In contrast with vision or audition, smell's physical stimulus has not been captured in a comprehensive classification. This failure is not due to the seemingly subjective nature of smells, but rather to the molecular complexity of the olfactory stimulus. Although it may be plausible to distinguish spatial features of visual objects, like shape, orientation, and movement, and to map them onto functional regions in the brain, it appears less intuitive to distinguish qualitative features of odor objects, such as apple, garlic, or urine, and to map them onto specialized brain regions. This is especially true considering that this qualitative space does not yet have clear links to physical structures.

Odor perception confronts us with many engaging perceptual experiences. For example, although we know that the perception of flavor takes place in the nose, we feel its location in the mouth. (Sensory scientists describe this as "oral referral.") Further, you might have encountered the "tip of the nose" phenomenon in which you are unable to name or describe a smell, even a familiar one, although you recognize it as such. ("Wait, it's on the tip of my tongue; I know this smell!") Moreover, odor perception is notoriously easy to manipulate through verbal and cross-modal cues. A wine expert may call out a vanilla note, which you had not smelled and only now seem to perceive. Besides, depending on the context, the perceptual quality of otherwise identical odorous mixtures can be experienced quite differently. (If we blindfolded you, could you tell the odor of stinky cheese apart from that of smelly feet?) We will look at these phenomena to explore what smells are and what sort of things or qualities they represent in the world. Or, in a scientific context: What kind of perceptual information is represented through odors, and how can that be linked to neural correlates in the brain?

A Modern Model for Sensory Neuroscience

The scientific history of smell is remarkably modern. Olfaction was catapulted into mainstream neuroscience almost overnight with the discovery of the olfactory receptor genes by Linda Buck and Richard Axel in 1991. Seldom has a scientific discovery shaped an entire field as profoundly as this one. Both Buck and Axel received the Nobel Prize in Physiology or Medicine for their work in 2004. It turned out that the olfactory receptors constitute the largest protein gene family in most mammalian genomes (except for that of dolphins), exhibiting a plethora of properties significant for structure-function analysis of protein behavior. The olfactory receptors were identified as belonging to a superfamily of proteins:

so-called G-protein coupled receptors, involved in all sorts of fundamental biological mechanisms ranging from vision to the regulation of immune responses. Finally, the receptor gene discovery provided targeted access to probe odor signaling in the brain.

What marks the olfactory receptor discovery as meaningful is the sheer force with which the genetic findings came to light. In one big step, this discovery overthrew a long list of unwarranted but century-old postulates about the sense of smell. Among them were the ideas that olfaction was not very sophisticated as a molecular system, that smell was in evolutionary decline in humans, and that the olfactory pathway presented a weird little system operated by causal principles different from those of the other senses. In sum, popular opinion had it that we had little to learn from smell and how it sheds light on how other sensory systems work. Nothing could be further from the truth!

The need to develop a theory of odor perception, which centers on the specifics of olfaction, has arisen quite recently. Technological advances shape modern science, which relies heavily on tools to visualize the unobservable and measure the uncertain. The principal reason smell had been absent almost entirely throughout the history of science is simple. It was, by its very nature, not easy to study. Imagine yourself being an experimenter in the nineteenth century. How would you make sense of smell? What kinds of tools could you use? How would you render the fleeting and transient appearance of odor visible, materialize its physical pathway, to trace its causes and experiment on smells in a controlled manner? Finally, how would you even begin to define and compare the perception of individual odors? The heart of this problem was expressed in 1914, in a challenge offered by none other than Alexander Graham Bell:

> Did you ever try to measure a smell? Can you tell whether one smell is just twice as strong as another? Can you measure the difference between one kind of smell and another? It is very

obvious that we have very many different kinds of smell, all the way from the odour of violets and roses up to asafoetida. But until you can measure their likeness and differences you can have no science of odour. If you are ambitious to found a new science, measure a smell.[1]

Before 1991, olfaction had not been very attractive as a domain of experimental inquiry: it did not promise glory in the annals of science. Curiosity and passion drove the scientists who worked on smell. Then, after the receptor discovery, the field started to open up to mainstream molecular biology and neuroscience. Suddenly, there was funding. Principal techniques could be applied, more detailed experiments were possible, and new people entered the community. The field began to change, and substantial shifts in disciplinary perspectives and progress began to unwind. Today we sit right at the heart of these developments.

We have learned more about the olfactory pathway over the past couple of decades than in all earlier centuries combined. But instead of decisive conclusions about the mechanisms of smell, we have arrived at much deeper questions; these questions now invite us to revisit the central assumptions about smell perception. This creates an intersection where science and philosophy interlace.

A Philosopher at the Laboratory Bench

How difficult can it be to crack the code in the nose, really? I still remember the day when I began to realize the true scope of this question. It was January 2014 when, with mounting excitement, I was traveling from Vienna to the United Kingdom, en route to meet my first olfactory scientist. It was promising to be an unusual meeting for many reasons. For one, Stuart Firestein is not your garden-variety scientist. Formerly a theater director in Philadelphia, he became a renowned neuroscientist prior to realizing an interest in the history of science. For another, back then, I was not yet a scientist. I had recently finished my doctorate about odor classification with a background in the history and philosophy of science. It all happened quickly. A few weeks previously, my dissertation examiner, the incomparable Hasok Chang, had contacted me out of the blue: "Do you know this guy called Stuart Firestein? He's here for a sabbatical at Cambridge, and I think you two should talk."

Little did we know that, about a year and a half later, I would find myself working in the Firestein laboratory at Columbia University. It was a match made in heaven, and I worked there for the next three years. Our first meeting ended about 5 a.m. and possibly would have continued for much longer if energy had been of endless supply. Over several pints of beer, it unfolded that the whole molecular affair of smell was not only much more complicated than previously imagined, but that it might even be different from how the olfactory system had been assumed to work for the last couple of decades! Things in the science of smell were moving fast, so fast that one could all of a sudden see a piece of scientific history unfolding, live stream, and in its own time. I was hooked.

It was an incredible opportunity to talk to the very people who were shaping a field, doing the crucial experiments, and working at the frontiers of an emerging breakthrough in the scientific understanding of an elemental question: What does the nose know?

To capture the spirit of the field in this monumental time, this book is framed around years of recorded conversations with past and present actors in the field. The nature of these meetings, covering hours and hours of discussion about olfaction and revolving around its past and modern challenges, ended up being different each time. They happened over coffee at conferences, with beers at the bar, perched on stools in the lab, or on the phone. Some talks lasted for about an hour; others continued over several days, couch crashing and long car rides inclusive. These recordings are informal testimonials to the intellectually dynamic and welcoming personality of the field. As Paul Breslin, of the Monell Chemical Senses Center, remarked to me during the fortieth meeting of the Association for Chemoreception Sciences (AChemS): "The field has no discipline. We are united by a common question."

Writing this book with the perspectives of the voices that created the content for its narrative cuts across the barrier separating published from practiced science.[2] It shows science in action. The neat public image of a hypothesis-driven approach to the scientific method cannot nearly capture the trials and errors, the cul-de-sacs, the creativity arising from failures and uncertainties, and the discussions involved in making an experiment work.[3] The speed at which modern science accelerates requires us to look more closely at the changeable nature of hypotheses, explanations, evidence, and accuracy.

And so, in addition to an inquiry into olfaction, this book also offers a reflection on science in practice. Science is highly pluralistic, and routinely more open-minded to critical dissent than someone outside the laboratory would imagine. Traditional understanding of the advancement of science builds on Thomas Kuhn's landmark 1962 book, *The Structure of Scientific Revolutions*. Its ideas influenced generations of scientists and philosophers, who have taken from Kuhn the view that everyday science operates as the activity of "filling the empirical gaps" within the prominent theory of its time—until a revolution occurs.

But what if we stand right at the forefront of a scientific field, amid its open questions and ongoing developments? By being in the middle of the laboratory trenches, we can observe normal science in another way, from a perspective that highlights its much more dynamic and curious nature. How does scientific progress look from an up-close vantage point at the experimental frontier? This was the question with which I had started this book. Like the science it engages with, this question soon started to evolve.

The making of *Smellosophy* led me to subscribe to the argument Patricia Churchland presented in her benchmark 1986 work *Neurophilosophy:* "if you want to understand the mind, you need to understand the brain."[4] Paul Churchland followed a similar program from a neurocomputational viewpoint, and John Bickle outlined this concept as psychoneural theories.[5] Twentieth-century neuroscience has revolutionized our understanding of the brain as the origin of the mind. It challenges many traditional philosophical intuitions about cognition and its architecture. Rather than accommodating scientific insights into our existing philosophical worldview, the reverse is required. What (potentially new) philosophical questions about mind and brain arise from the impressive progress in neuroscience? In this context, the sense of smell presents an excellent opportunity to put such a philosophical program further to the test.

From the Air to the Brain and into the Mind

In ten chapters, this book outlines an integrated perspective for our study of olfaction that is also applicable to theorizing about perception more generally.

For its narrative to unwind, *Smellosophy* advances through (roughly) four themes: history, philosophy, neuroscience, and psychology. Seldom combined, each approach holds a piece to the puzzle: psychological phenomena are expressions of neural processes, and a synthesis of their explanations benefits from a philosophical angle that's been informed also by the history of an inquiry.

First, we uncover the historical development of scientific interest in the sense of smell. Chapter 1, "History of the Nose," engages with the distant past, from the ancient philosophers to the emergence of an olfactory research community in the mid-twentieth century. These early, almost forgotten experimental records of research on smell offer an intriguing, yet untold, history of creativity in scientific reasoning. They also highlight, by omission, the key element that theorizing about olfaction must build on: an understanding of the biology of the sensory pathway. Chapter 2, "Modern Olfaction: At the Crossroads," explores how olfaction became part of mainstream neuroscience with the discovery of the olfactory receptors. It situates our current scientific understanding of smell and the olfactory pathway in the general context of brain research with its central paradigm of vision. Here, a challenge emerges that remains unanswered: Does it make sense to model olfaction via a mapping of odors onto neural structures?

This requires an answer to the fundamental question of what odors actually are, which leads us to a more philosophical stance regarding the characteristics of olfactory perception. In the following chapters, we see that smell is of much higher cognitive sophistication and behavioral relevance than is commonly believed. Chapter 3, "Minding the Nose: Odors in Cognition," investigates the role odors play as elements of the mind. Smell sits at the border of conscious and unconscious perception. So, how should we think of smells as mental objects? Does smell communicate conceptual content? If so, what kinds of things in the world do odors represent? Subsequently, Chapter 4, "How Behavior Senses Chemistry: The Affective Nature of Smell," looks at the perception of odor from a complementary biological and social angle. Olfaction has a powerfully affective dimension. Smells can influence our mood and evoke physiological reactions as well as emotional reactions; some olfactory experiences create strong bonds with distinct memories. Odor perception plays multiple roles, including roles in human behavior. How can these various kinds of effects be explained? Chapter 5, "On Air: From the Nose to the Brain," explores the different ways in which an answer to this question must be sought. It centers on the case of spatial navigation with odors to specify explications involving the chemical stimulus, embodied processes (such as sniffing), neural topology, and perceptual space. To understand how odors are realized as mental sensations necessitates a closer look at the neural topology of the olfactory system. Chapters 6 to 8 discuss smell from a neuroscientific view. We journey through the structure of the olfactory pathway from the olfactory receptors, via the olfactory bulb, straight into the olfactory cortex where the heart of the argument unfolds—namely, why olfaction challenges current conceptual foundations of neuroscience based on the visual paradigm. The olfactory signal gets scrambled, so much so that the neural topology bears absolutely no resemblance to the topology of the chemical stimulus. The idea that our mental life is, in one way or another, a representational expression of physical structures breaks down. Perception does not mirror the world; it interprets it. Specifically, Chapter 6, "Molecules to Perception," demonstrates how stimulus-response models that exclude receptor mechanisms are doomed to fail. This happens for two causal reasons: On the one hand, olfactory receptors do not respond to molecular features as dictated by the principles of organic chemistry. On the other hand, two molecular mechanisms at the receptor level dismantle the signal, leading to a case of severe sensory underdetermination. Chapter 7, "Fingerprinting the Bulb," counters the common belief that the olfactory bulb results in a chemotopic map, meaning a fixed systematic representation of chemical stimulus features. It breaks down the widespread idea that the bulb's function is spatially determined by looking at its developmental origins. Chapter 8, "Beyond Mapping, to Measuring Smells," advances an alternative to the traditional understanding of primary sensory cortices, tackling the following question: If the brain does not process odor signals as a stimulus map, how else should olfaction be modeled? In exploring the piriform cortex, and its high connectivity with neighboring cortical domains, one alternative is to look at the temporal features of neural activity in odor categorization. The olfactory brain, this part concludes, is much closer to a measuring instrument than a map.

Chapter 9, "Perception as a Skill," reveals how these recent neuroscientific insights mirror psychological explanations about perceptual learning and expertise. It shows that variability is inherent in the process of olfactory perception. The fact that people perceive and describe smells differently is not merely a matter of subjective experience. Subjectivity implies an absence of objective measures. Perceptual variation in olfaction, however, resides in definitive causal processes, including receptor genetics and the highly distributed neural representation of the olfactory system. Your nose is tailored to measure the world as calibrated by your mental life and physiological conditions. These processes have an objective basis, analyzed with the case of sensory experts in perfumery and winemaking. Apparent idiosyncrasies of olfactory perception are readily explained through observational refinement and skill building, drawing on several distinct cognitive mechanisms. Last, Chapter 10, "The Distillate: The Nose as a Window into Mind and Brain," situates olfaction in the bigger picture, with general theorizing about perception and the brain.

How the brain guides the nose to interpret its information and makes better choices is an intriguing and highly complex question. For this reason, even though this book shows science presently in action, *Smellosophy* is also a book about the future. It will end where new prospects for the science and philosophy of smell rise.

That the modern science of scent is heading fast toward the future becomes clear once we compare it with how the nose was studied in history.

CHAPTER 1

History of the Nose

If all things were turned to smoke,
the nostrils would distinguish them.

—Heraclitus

The scientific biography of olfaction can be summarized in one sentence: Odors have always presented an ontological problem. Smells have been primarily of interest in explanations of other phenomena, such as animal behavior or flower pollination. The characteristics of smells *qua* smells, however, have received little consideration. The features of olfaction are challenging to define and measure. Scientific experimentation surrounding the sense of smell has never arrived at a resolute answer.

The historical record reveals what has been missing in our attempts to understand olfaction. It is a display of omission. What remains wanting is an organismal approach: a systematic investigation starting from the sensory system that gives smell meaning. Historical explorations of smell perpetually centered on the objects that emanate odors as their material basis. It still is common today to start from the properties of the physical stimulus of smell to arrive at a definition of its perceptual effects. This strategy has been successful in research on other senses; its adoption in olfaction was only intuitive. Past patterns notwithstanding, it may be time to revisit the real advances of this approach and reevaluate the assumptions underlying such object-centered views on olfaction by reconsidering the conditions of contemporary knowledge via its historical emergence. These records of historical research on smell also offer an intriguing, untold history of creativity in scientific reasoning.

Ancient Times

Most scientific histories start with the ancient Greeks, and this story is no exception. Long before the Renaissance, the cradle of modern science, theories of smell went back as far as Aristotle and Plato. According to premodern belief, everything in the universe was composed of four elements: earth, water, air, and fire. This theory, championed by Empedocles, explained the order of things. Each element was assigned its natural place in the greater cosmos, and these elemental qualities determined everything else in the world. Fire was hot and dry, air was hot and wet, water was cold and wet, and earth was cold and dry.

How did odors fit into this worldview? Smells were observed to travel great distances from their sources to the perceiver. Vultures, for example, could find cadavers to prey on over considerably long distances. Nonetheless, it was not evident what kind of matter allowed for such effective transmission. Would odors require a medium? Given its indeterminate and insubstantial nature, smell in humans was not considered of great importance by ancient philosophers, and its discussion did not occupy a critical percentage of their works. Most thinkers proceeded from the common belief that smell was dominant in animal life but auxiliary in humans. Their sparse reflections about the causes of odor coalesced around two opposed conceptualizations of materiality: odors as particles versus odors as waves.

The first ideas about smell were atomistic. Democritus and the later Roman philosopher Lucretius speculated about pleasantness.

They thought that pleasant smells were caused by round particles, while unpleasant ones had edgy shapes like triangles. An immediate obstacle to this idea was the lack of explanation of how the same odorous liquid could cause different qualities and varying degrees of potency. Plato avoided this dilemma. He thought smells emerged from the physical movement of fine particles. None of the four elements were odorous themselves, so these qualities had to arise through the material conversion of elements into vapors or fumes in transitions of water and air. In the dialogue *Timaeus,* Plato analyzed odors in an intermediate, hybrid character that eluded concreteness. Their chimeric constitution prevented smells from forming natural kinds so that they could be characterized only by their hedonic appeal of pleasantness. What remains remarkable about Plato's idea is its implicit understanding of smell as a sign of material change. Emphasis on smells as expressions of elemental transitions suggested a form of sensory judgment about the contingent and mutable nature of matter.

Aristotle opposed the vapor theory, pointing out that fish detect odors in water. Aristotle did not criticize his teacher Plato openly but ambushed Heraclitus, who held similar views. In *De sensu et sensibili,* Aristotle argued for the necessity of a medium and a wave theory of odor. This move was based on Aristotle's metaphysical worldview that led him to reject the possibility of a vacuum (*horror vacui*). As waves, odors conveyed information by acting as a formal cause on the medium of air or water. In Aristotle's metaphysics, each object presented a compound of two causes: matter (*hulê*) and form (*morphê*), and their interaction described the theory of hylomorphism. Qualitative variations in odors, on this account, resided in differences in the material composition of waves.

Two ideas of Aristotle remain notable. The first is the distinction between two types of odor in human olfaction. In *De sensu,*

Aristotle recognized that some smells presented accidental properties. Their hedonic qualities were not intrinsic to their object but were dependent on the observer; for example, food aromas became more pleasant in a state of hunger. Other smells were intrinsically pleasant or unpleasant, regardless of the observer's constitution. These smells were substantial properties without interests to human desire; this second type of odor appeared to be aesthetic, such as the fragrance of flowers.

Second, Aristotle alluded to a cognitive layer in the perception of odor and flavor. Odor and flavor brought "into actual exercise the perceptive faculty which pre-existed only in potency. The activity of sense-perception thus is analogous not to the process of acquiring knowledge, but to that of *exercising knowledge already acquired*."[1] Perhaps it was their link to personal memory that rendered smells unsuitable for philosophical inquiry into the fundamentals of reality.

Smells were treated as objective properties in horticultural and medical practice. Theophrastus, student of Aristotle, examined smells by therapeutic powers. His *Enquiry into Plants and Minor Works of Odours and Weather Signs* provided systematic advice on the treatment of ailments with the odorous sap and juices of plants, such as the use of cabbage to counteract the effects of wine and "expel the fumes of drunkenness."[2] Theophrastus noted eight kinds of flavored fluids: sweet, oily, sour, astringent, pungent, salty, bitter, and acid. Some of these types formed further subgroups: sweet juices encompassed four varieties.

Theophrastus further described the varying material effects of odors. As dry elements, odors emerged after the evaporation of moist elements. Flavors in plant juices were a mixture of dry (earth) and moist (water) components. This separate constitution explained how plants could differ in smell and taste—for example, why some fragrant substances tasted bitter while sweet substances were frequently odorless.

The key to their medicinal use was that, next to dosage, certain odors and flavors were beneficial to some organisms and detrimental to others, depending on their constitution. This conflicted with the doctrine of Aristotle, who thought that a pair of opposites characterized any substance. In Aristotle's view, odors were forming either good "natural kinds" (such as sweet) or represented harmful privations, thereby constituting a negation of their positive counterpart (like bitter as opposed to sweet). But Theophrastus considered the effects of smells as coupled to the perceiving organism and in parallel with the elementary composition of plants.

Variation in the perception of odor, therefore, was linked to the constitution of the observer. Support for this compositional theory of odor came from observations that some qualities (like pungency or sweetness) could be attributed to pleasant- and unpleasant-smelling substances alike. Theophrastus's ideas may count as an early foray into blended sensations. His treatment of odors in *De causis plantarum* (*De odoribus*, or fragment VI) is extensive yet disorganized. No proper theory of odor emerged from these ideas.

Medieval Times

Throughout the Middle Ages, when medieval society matched sins to the senses, smells acquired a notorious reputation.[3] Odors acted as moral signatures, revealing the true essence of things. Numerous treatises elaborated on the divine order, contrasting pleasant smell and moral virtue with foul odor and vice. The sulfurous fumes of Hell, the emissions of decay from rotten flesh and fruits, or the stench of disease as a punishment for sinful behavior were in belligerent opposition to the scent of flowers as a mark of beauty in God's creation, where visions of paradise were populated by blossoms and prayer rituals accompanied by incense. Even in death, the "odor of sanctity" revealed a saint or martyr. Instead of the reeking trail of a rotten corpse, the body or tomb of a holy person emanated a pleasant mellifluous fragrance, like honey or sweet flowers and herbs.

Odors embodied tangible signals of spiritual order, an expression of natural rule. Theological examination was inseparable from early, prescientific inquiry into the cosmos and, in this context, smell expressed a pervasive fabric of reality in medieval thinking. As gateways between the corporeal and the immaterial, odors reached beyond their immediate physical environments by communicating a world of concealed meanings and causes.

As expressions of intrinsic essences, odors proved an invaluable tool for the physician. Medieval medicine drew primarily on insights of the humoral theory of the Greek physiologist Galen. Health and disease were expressions of the relationship between four bodily fluids (blood, phlegm, yellow bile, and black bile) that corresponded to the four elements (air, water, fire, and earth, respectively). Each pair was assigned a particular quality. Their balance was responsible for both the temper and the physiology of an individual. Blood stood for courage, phlegm for a calm temperament, yellow bile for an ambitious or restless persona, and black bile for analytical faculties. The body constituted a system of natural signs, giving clues for inferences about unobservable psychological phenomena. Smell was a critical part of this notion. Fragrances similar to one's humoral composition were experienced as pleasant; diverging ones were disagreeable.

Physicians saw odors as diagnostic tools and treatments to rebalance humoral fluids and tempers. The poplar device of "urine wheels" classified the maladies of patients according to differences in the smell, color, and—yes—taste of their excretions. Like other medical inventions, this practice attracted charlatans, who looked at people's urine for purposes of divination. These charlatans became aptly known as "piss prophets," a term documented as early as 1655.[4]

Medieval interest in the transmission of smells was limited and part of general discussion of ancient cosmology, with the majority of scholars favoring Aristotle's medium over Plato's vapors. Supporting Aristotle's theory was the circular expansion of odors and the fact that cadavers could attract predators over great distances in the absence of wind. The Andalusian polymath Averroes (Ibn Rushd) also mentioned that bees detected smells without inhalation. Averroes distinguished two types of essences in sensory perception: a material essence (*esse corporale*) and an immaterial essence (*esse spirituale*); the former was grasped through the sensory organs, the latter perceived by the soul.[5]

Most scholastics, including Thomas Aquinas and Petrus Hispanus (author of *Scientia libri de anima*), believed in a wave theory of odor. Medieval scholarship shared Aristotle's fear of the vacuum. Only the Dominican monk and physician Albertus Magnus attempted to reconcile the contrasting views on odor by Aristotle and Plato. In *De anima*, Magnus hypothesized that Plato's particles were the real cause of smells before they were transformed into a spiritual quality through a pneumatic medium in the sensory pathway. He linked his argument to observations of particles in poisonous fumes. Magnus advocated a multistage theory of perception: sensory qualities were abstracted from their material form through an intellectual realization in the sensory organs.[6] The idea of a medium prevailed, however, in theories of odor.

Just how waves acted as a medium was in dispute. The Persian physician Ibn Sina, known as Avicenna, offered three interpretations. Maybe particles intermingled with a medium (air or water). Alternatively, odors might arise from material changes in the medium. Or perhaps the medium transported smells as informational waves. Avicenna arranged smell sensations by their pleasantness (next to a scale of sweet and sour). More unusual was his idea that, in addition to *sensibilia*, odors also communicated primary properties like form, number, and movement or rest.[7]

But what about physiology? Confusion surrounded medieval understanding of the olfactory pathway, especially the olfactory nerves. After their discovery by Galen in ancient times, olfactory nerves were initially excluded from the group of cranial nerves.

This view was corrected only in the seventeenth century by Caspar Bartholin the Elder.[8] Medieval theories about the physiology of smell were scant, with one notable exception: Bartholomeus Anglicus (Bartholomew the Englishman), a thirteenth-century Franciscan monk. Bartholomeus reviewed anatomical knowledge of the olfactory pathway. He corrected observations such as those by Constantine the African (a physician in the eleventh century who was known for his translations of Arabic medical texts), who thought that the function of one nostril was to draw in air to the brain while the other ejected excesses. Bartholomeus asserted that the nostrils had sinus tissue protruding like teats from the brain, transforming the inhaled dry fumes into soulful spirits. Bartholomeus further hypothesized that smells act directly on the brain. As the present-day historian Chris Woolgar explains, "The nose was a passage to carry smell into the brain; the bone separating the nose from the brain was porous and the air passed through for smell to be sensed by two projections from the front ventricles of the brain" so that "in the case of smell, the brain itself was a sense organ."[9]

Soon enough, the medieval world began to fade and the ancient theory of the four elements broke down. With the Copernican revolution in physics, the universe was no longer finite. Replacing previous ideas of a closed world, the idea of an infinite universe entered. Consequently, new challenges opened up for students of nature in the Renaissance and the early modern period. The cosmos and its objects lost their natural place, now indeterminate and indefinite in its constitution. Once ancient ontology was overthrown, old ideas started to mingle with the new. Timeworn questions about how many things there are, and to what kinds they belong, were reinvigorated. Without such a fixed metaphysical landscape defining the elementary basis of things, what were odors to become?

The Modern Period

Smells experienced a surge in scientific significance throughout the rise of eighteenth-century botany. Central to this development was the father of modern taxonomy, the Swedish scientist Carl von Linné, better known as Linnaeus. He developed an excessive fascination with grouping and ranking everything that passed his way: plants, animals, minerals, and even his colleagues. The success of the Linnaean system of classification was its hierarchical integration of individually diverse elements into general categories. The invention of a binominal nomenclature, rigorously dividing items into genus and species, proved ingenious. Driving this work was Linnaeus's ambition to devise a system that could capture and sort everything there is. He almost succeeded. Only smell disobeyed his order.

Linnaeus examined therapeutic purposes of plants based on the affective nature of their fragrance. His dissertation, *Odores medicamentorum,* was coauthored with his student Andreas Wåhlin.[10] It is referenced as the first systematic arrangement of smells. It did not offer a classification of odor, however, but discussed odors as indicators of medicinal powers in plants. Linnaeus's systematization of odor deviated considerably from his general taxonomic principles; the reason for this methodological exception remains unknown. Instead of a nested hierarchy, Linneaus divided odor sensations into seven classes along a gradual scale of hedonic appeal (invoking different degrees of pleasantness). Pleasant smells encompassed categories such as aromatic or fragrant, like warm clove or lily. Unpleasant smells included ambrosial (musky), hircine (goat-like), foul (repulsive), and nauseating (disgusting) odors. An exception was the seventh class of alliaceous (garlicky) odor because it did not unambiguously fit this hedonic scaling.

Linnaeus published a refined schema in 1766, the *Clavis medicinae duplex* (*The Two Keys of Medicine*). Aimed at therapeutic use, "ways of life" were correlated with "properties of nature." Linnaeus assigned five principles of antithetic effects on a hedonic scale of sweet-smelling versus evil-smelling scents (after Hogg's translation).[11] Interest in the affective nature of smells also concerned their behavioral impact. The Swiss anatomist and father of modern physiology Albrecht von Haller was intrigued by the changeability of odor experience over time. For example, smells like fresh musk or civet began with a highly unpleasant quality of a fecal-like odor before they acquired more pleasant notes. Also, throughout different stages of decomposition, decaying materials exhibited a series of shifts in their olfactory quality, from putrefied to sweetness. Virtually infinite ways to split up the realm of odor were conceivable. In *Elementa physiologicae* Haller thus organized smell by three generalizable principles of appeal: sweet-smelling or ambrosiac, intermediate, and stenches.[12] These categories subsumed a miscellaneous collection of odor qualities, ranging from saffron to feces to "the odor of bedbugs found in coriander."

Sweet-Smelling	*opposed to*	*Evil-Smelling*
AMBROSIAC		RANK-SMELLING
dilating	Libido	*Aphrodisiac*
Splitting		inflating
FRAGRANT		REEKING
sedative	Sleep	*Soothing*
Inebriating		invigorating
SWEET-SCENTED		STINKING
inciting	Vitality	*Analeptic*
Anaesthetic		stifling
AROMATIC		NAUSEATING
stimulating	Activity	*Warming*
Evacuatory		spasmodic
ORGASTIC		PUNGENT
astringent	Consciousness	*Compulsive*
Stupefying		relaxing

Linnaeus and Haller incited further interest in olfaction, with the Dutch physiologist Hendrik Zwaardemaker leading the way. His 1895 manuscript, *Die Physiologie des Geruchs* (*The physiology of smell*), was a comprehensive survey of olfactory theories by predecessors and contemporaries.[13] Zwaardemaker distinguished three types of odor sensation: pure odors (*reine olfactive Riechstoffe*) and two kinds of mixed sensations—namely, olfactory impressions accompanied either by pain in the nose (*scharfe Riechstoffe*) or flavors in the mouth (*schmeckbare Riechstoffe*). Zwaardemaker thought that the problem with research on pure odors was the lack of proper names (*keine besonderen Namen*); odors were commonly identified by their material origins. He compared the underlying dilemma with prescientific descriptions of color. Before Newton, colors had been determined by reference to paradigmatic objects, like blood for red. Methodical terminology for color vision developed after the discovery of the light spectrum.

Scientific classification of smell thus required the successful isolation of simple smells, analogous to primary colors. The isolation of odor components posed tremendous technical difficulties at the time. Slight contaminations and different concentrations of stimulus samples had notable effects on olfactory quality. Zwaardemaker was forced to choose an alternative approach. He outlined nine primary odor classes, which served to integrate botanical, chemical, and physiological perspectives on fragrant materials:

I. *Odores aetherei* (etherous odors; *ätherische Gerüche*)

II. *Odores aromatici* (aromatic odors; *aromatische Gerüche*)

III. *Odores fragrantes* (balsamic odors; *balsamische Gerüche*)

IV. *Odores ambrosiaci* (amber-musk-odors; *Amber-Moschus-Gerüche*)

V. *Odores alicacei* (alliceous odors; *Allyl-Cacodyl-Gerüche*)

VI. *Odores empyreumatici* (burnt odors; *brenzliche Gerüche*)

VII. *Odores hircine* (caprylic odors; *Caprylgerüche*)
VIII. *Odores tetra* (repelling odors; *widerliche Gerüche*)
 IX. *Odores nausei* (vomit-inducing or fetid odors; *Erbrechen erregende oder ekelhafte Gerüche*)

Fragrant materials eluded consistent and comprehensive systematization. Even some of Zwaardemaker's observations stood in stark contrast with his system, such as the case of heated arsenic smelling similar to garlic. Additionally, the classification of aromatic materials became more difficult with the discovery of synthetics in late nineteenth-century chemistry.

Botanists continued to sort odorous materials despite the futility of mastering the sheer diversity of fragrant substances. The Austrian botanist Anton Kerner von Marilaun found a chemical understanding of smells necessary yet insufficient. Smell engrossed complex meanings in the living world, irreducible to chemical formulas. A case in point was evidence of olfactory mimicry, in which plants emanate smells resembling odors from other species and taxa that scam unsuspecting insects into pollination. He published his observations in *The Natural History of Plants, Their Forms, Growth, Reproduction, and Distribution.*[14]

In von Marilaun's theory, two core purposes, survival and reproduction, defined odor either as an attractant or a repellent. Merging chemistry with biology, his system covered five primary chemical groups: indoloid, aminoid, paraffinoid, benzoloid, and terpenoid scents. Four features specified the function of these groups: descriptions of odor quality, chemical composition, botanic origins, and value. Even in von Marilaun's view, this schema remained fallible. Most smells were not simple but complex chemical mixtures. Besides, plants could emanate various fragrances during their development, or according to diurnal and annual cycles. The biological basis of odor showed not categorical, but overlapping distinctions.

Still, von Marilaun's classification remained popular in horticulture. A keen proponent of von Marilaun was the botanist John Harvey Lovell, who was commissioned to write a series of seven articles on flower smells for the *American Bee Journal* in the 1920s. This series of articles spanned a range of diverse issues, including a general physiological introduction to human olfaction and its relation to taste, a classification of flower odors, and a systematic survey of how flower scents affect bee behavior. Notable about Lovell's work is his second essay. In *Classification of Flower Odors,* he remarked on the arbitrariness of any classificatory system as it is bound to practical applications. Lovell also found that odors invited a certain degree of ambiguity in their classification: "In many instances there will be a difference of opinion as to the odor of certain flowers." Such divided opinion is likely because "a flower may exhale two odors, or its odor in the morning may be different from that in the evening."[15]

Another advocate for von Marilaun was Frank Anthony Hampton. In *The Scent of Flowers and Leaves,* Hampton offered ten categories and presented three distinct reference standards to identify known smells and accommodate novel ones.[16] The first standard was a distinct odor quality (verbal descriptors), the second was its main utility for producing fragrant materials, such as essential and fatty oils or alcohols (extracted plant substances), and the third was flower type (specimen).

Taxonomic interest in odor classification dwindled by the mid-twentieth century. Fundamental changes in the biological sciences during the nineteenth and especially the twentieth century accelerated this tendency. Technological breakthroughs and central insights into biochemical processes shifted scientific focus. Natural history made way for genetic and experimentally driven inquiry into animal and plant life. Odors lost explanatory value in this new understanding of the life sciences.

The Chemical Turn

Prior to the nineteenth century, few scientists examined smells as chemical entities—aside from in perfumery. One of the two oldest professions in the world, perfumery was a profoundly secretive trade. The history of perfumery remains full of untold tales, and its intricate connection with early chemistry seems hard to disentangle. In its principal methods, perfumery represented a practice of early chemistry. Chemistry and perfumery significantly overlapped in their use of materials, instruments, and objectives.[17]

Perfumers extracted, distilled, mixed, heated, separated, and experimented with the observable characteristics of fragrant materials. Documentation of techniques for creating and manipulating fragrant materials go well back to the pre-Christian era. The first chronicles detailing the use of oils and pomades reach as far back as ancient Egypt.

Over centuries, perfumers developed and perfected several techniques. In the process of *extraction,* mechanical force was applied to plant materials by pressing or grinding. Substances processed by this method were usually rich in volatile essential oils and cheap to farm, like the peel of oranges. Another technique was *distillation,* referring to either dry or moist steaming. Here, materials like flowers, or wood, were exposed to heat and their fragrant extract collected through condensation. Some flowers, such as jasmine, would denature in a distillation process. More delicate materials underwent different processing via *maceration,* an operation involving the use of solvents, such as spirits, to separate particular components. Extremely delicate flowers were subordinated to *enfleurage* or absorption. Flowers were spread over a frame that contained a layer of fat absorbing their odor within a period of seventy-two hours. This was an expensive and time-consuming method. The choice of procedure could involve the price of the products, the desired quality of the processed substances, and their final application as essences, waters, oils, pomades, or balms. These techniques characterized the practice of perfumery for centuries, until the fourteenth century.

Around 1320, a hallmark invention by two Italians marked the beginnings of modern perfumery: the serpentine cooling system facilitated the production of high-grade alcohol. It opened unforeseen opportunities to the perfumer. With the production of high-grade alcohol, the application of scented products changed radically, as alcohol dilutes and breaks down mixtures. Perfume ingredients could be separated and released throughout several temporal stages, and fragrant creations manifested different qualities over time as a function of their duration of contact with the skin.

Modern perfume was born. It transformed into a threefold composition entailing a "top note" (released within the first fifteen minutes), a "heart note" (spanning about thirty minutes after the top note evaporated), and a "base note" (the remainder, lasting up to twenty-four hours). During the Renaissance, there was an obsession with creating new and complex kinds of fragrances. Hungarian Water, commissioned by Queen Elizabeth of Hungary, was one of the first alcohol-based perfumes, created in 1370. It remains one of the most successful fragrances.[18]

Little was known about the chemical components of smell. The Anglo-Irish Robert Boyle, trailblazer of modern chemistry, changed that. In 1675, Boyle conducted a series of twelve *Experiments and Observations about the Mechanical Production of Odours*.[19] These short reports, instructions for experimental reproduction, constituted part of Boyle's broader criticism against the popular chymist (precursor of modern chemistry) doctrine by his contemporaries called the *tria prima*, advocated by Paracelsus and his followers, the Spagyrists. The tria prima outlined the composition of matter according to three principles: salt as the principle of fixity and incombustibility, sulfur as the principle of flammability, and mercury as the principle of fusibility and volatility.[20]

Boyle saw the chemical world as composed of particles (corpuscles). Smells were no exception. Odors, however, presented a source of uncertainty. It was evident that materials emanated all sorts of smells; less obvious was their "smelling principle." The problem for a corpuscular theory of smell was that, despite continuously giving off odorous particles, their sources did not seem to lose significant amounts of weight. Observing a piece of asafoetida for six days, Boyle speculated: "The whole lump had not lost half a quarter of a groin; which included me thinking, that there may perhaps be streams discernible even by our nostrils, that are far more subtil than the odorous exhalations of spices themselves."[21]

Did changes in olfactory qualities correspond with different chemical reactions? Boyle devised several trials with variations, testing for the impact of dilution, heat, or the use of different metals for vessels, such as silver or gold. The reactions were distinct and measurable, albeit diverse. For example, Boyle found that the combination of some odorless materials produced a strong-smelling odor. At other times, he obtained a pleasant smell from stinking ingredients. Additionally, it was possible to neutralize, or even enhance, some odors through the addition of barely odorous substances. These experiments demonstrated that the production of odors obeyed the same laws as other chemical reactions. By way of an example, one of his experimental instructions in his *Experiments and Observations* reads:

EXPER. I.
With two bodies, neither of them odourous, to produce immediately a strong Urinous smell.

Take good Quick-lime and Sal Armoniac, and rub or grind them well together, and holding your nose to the mixture, you will be saluted with an Urinous smell produced by the particles of the volatile Salt, untied by this operation, which will also invade your eyes, and make them to water.

The corpuscle view of odor became scientific mainstream by the eighteenth century. But the smell of these airborne particles was not explained in purely physical terms. An immaterial and animated essence was assumed to be involved in the perception of smells. The most articulate formulation of this idea was the theory of *spiritus rector;* its leading proponent was the Dutch botanist, chemist, and physician Herman Boerhaave, von Haller's teacher.

Boerhaave assumed two elements responsible for odor sensations, separating the causality of physical matter from their mental experience. The effective material causing the transmission of odors was volatile particles. A pure particle view could not explain the variety of odor quality, however, since these particles were homogeneous. The qualitative dimension of smell was the spiritus rector, an invisible oily substance appending the physical particles and acting directly as some form of a vital force on the mind of the perceiver. Boerhaave noted: "But the oily Parts are in some measure subservient to this Sense, since they fly off together with the spirituous Rector, and adhering to the Surface of the olfactory Membrane, render the Effect or Action of the odoriferous Particles more permanent and lasting."[22] Smell remained an essential expression of the living world that transcended merely mechanical causes.

Modern understanding of odor began when two French scientists took a closer look at horse urine. Horse urine, readily available and abundant in eighteenth-century Europe, possessed many distinct characteristics of experimental interest (bright color, alkaline properties, and pungent smell). Next to Lavoisier, these two scientists were the most well-known French chemists of their time: Antoine François, comte de Fourcroy (ominously linked to Lavoisier's untimely death), and Claude Louis Berthollet. They identified and isolated urea as the cause of the urinous smell of pee.[23] Others confirmed this breakthrough:

> The urine, after it has become alkaline, is often so tenacious and viscid that it can be drawn up in long threads. The

microscopic examination of the urine of the horse exhibits a great number of rounded corpuscles, from the size of mucus-corpuscles to four times that size, which burst upon pressure of the glass slips between which the fluid is examined. Four-croy and Vauquelin, after evaporating the urine of the horse, separating the urea as a nitrate, and neutralizing the acid by an alkali, found a small quantity of reddish fat, which volatil-izes over the water-bath, and is considered to be the cause of the smell and colour of the urine.[24]

In 1828, the German scientist Friedrich Wöhler catapulted smell into core chemistry with a follow-up experiment.[25] Wöhler synthe-sized urea from ammonium cyanate (CH_4N_2O). The importance of this synthesis is impossible to overstate. Organic matter, at the time, was thought irreducible to the principles governing inorganic matter. Organics obeyed separate laws and vital forces. Wöhler showed this not to be the case. He synthesized an organic sub-stance, urea, from an inorganic material, ammonium cyanate. Organic and inorganic chemistry were united—a paradigm shift in chemistry. For the science of smell, it was the starting signal.

A new material dimension of odors came to light. Piece by piece, the chemical composition of fragrance materials was identified, and active exploration of the synthetic reproduction of raw and rare materials began. In 1818, Jacques-Julien Houtou de Labil-lardiére determined that turpentine oil was composed of "a rela-tion of five C- to eight H-atoms $((C_5H_8)_x)$."[26] This discovery spurred analysis into the composition of similar essential oils. In 1833, Jean-Baptiste Dumas recognized that most essential oils exhibited notable similarities in chemical composition.[27] He divided essen-tial oils into "those containing only hydrocarbons such as turpen-tine and citron oil, those containing oxygenated compounds such as camphor and anise oil, and those with sulfur (mustard oil) or nitrogen compounds (oil of bitter almonds)."[28] Eugène-Melchior

Péligot, Justus Liebig, and Otto Wallach accumulated further insights into the constituents and formulas of essential oils important for perfumery, such as menthol and almonds. These discoveries went hand in hand with the improvement of techniques for the separation of different odor components from raw materials; these techniques included vacuum-distillation and derivatization techniques, producing structurally similar derivates from a particular chemical compound.

Over the following five decades, research on synthetics exploded. The synthesis of coumarin, in particular, acted as a catalyst for this development. Coumarin, first synthesized in 1868, with the smell of freshly mown hay, is naturally found in the tonka bean (*Dipteryx odorata*) and melilot or sweet clover (*Melilotus*). Utilizing the so-called Perkins condensation, coumarin was obtained from the condensation of salicylaldehyde (C_6H_4CHO-2-OH) and acetic anhydride (($CH_3CO)_2O$). Sir William Henry Perkins, for whom this reaction is named, was also responsible for the first synthetic dye, aniline dye—known as mauve today.

Marking the rise of fragrance and flavor chemistry was the synthesis of vanillin from coniferyl alcohol by Ferdinand Tiemann and Wilhelm Haarmann in 1874. Haarmann realized that academic interest in synthetics intersected with the growing demands of industrialization. He and Tiemann founded their own company shortly after (Haarmann's Vanillinfabrik). In the following years, reaction procedures were enhanced to accommodate the large-scale production of synthetics.[29] Haarmann hired Karl Reimer, who designed a technique for improving the synthesis of vanillin. Reimer's method was a complete success.[30] The company, renamed Haarmann & Reimer, grew vastly. (Much later, this firm was to become the fourth largest fragrance company, Symrise, after merging with another company, Dragoco.)

It was a time of fundamental change in the chemical production of fragrances and flavors, as well as dyes and inks. Several companies specializing in the production of synthetic materials sprang to life, including the two global players commanding the industrial fragrance market today: Firmenich (originally Chuit & Naef) and Givaudan, both founded in 1895.[31]

The industrialization of nineteenth-century Europe irrevocably shaped the face of modern chemistry. A lucrative market with synthetic aromas began. Higher production rates and demands for scented products fueled the hunt for more, better, and novel synthetic compounds, making the modernization of the food industry and perfumery inevitable. Raw materials traditionally used in perfumes—for example, ambergris—became too rare and expensive for wider commercial distribution.[32] Synthetic materials replaced raw materials. Synthetics are easier to handle in critical respects: they do not rely on seasons, like the farming of flowers does, and they are available at all times. Another factor was the introduction of ethical, hygienic, and legal restrictions regarding the use of animal products, like ambergris or civet, and their replacement with synthetics.

Synthetic chemistry spearheaded a fundamental shift in scientific understanding of odor. The Hungarian Leopold Ružička, 1939 Nobel laureate in chemistry for his research on insect pheromones, propelled insight into the binding capacities of molecules. As early as 1920, Ružička recognized the possibility that the osmophoric group might be responsible for the orientation of the molecule within a hypothetical receptor site.[33] (Ružička's career was exemplary for many flavor chemists at the time. Lacking support by traditional academic structures, Ružička turned to industry and became head of research and development at Firmenich.)[34]

The rise of chemistry altered the ontology of odor. It became uncertain what counted as natural or artificial. Chemical synthesis was more than an alteration of matter. It created a new perspective on the causal relationship between odors and their material basis. The invisible, intricately structured molecular world had surpassed the unruly order of botanic materials. The reality of odors, shaped by the development and evolution of botanic and animal beings, now was on a par with that of chemical scents artificially created in the laboratory five minutes ago. This ontological revolution revealed what was missing in previous approaches to olfaction—the sensory system shadowing the essential question: What does it mean to perceive smell?

Physiology at the Evening of the Nineteenth Century

Prior to the twentieth century, scientific interest in smell focused on the materials that emanate odors. When did we learn about the sensory system, its physiology and psychology? At the turn of the twentieth century, targeted scientific attention to olfaction emerged independently in diverse areas. And so our historical picture turns from chronology toward a mosaic of investigations, spanning early physiology, psychology, and, soon enough, biochemistry.

Initial interest in olfactory physiology was brief but creative. Little was known about the olfactory pathway to begin with. Hippocrates and Leonardo da Vinci had provided basic anatomical drawings of the nasal cavity. Detailed descriptions came later with the studies of the British surgeon Nathaniel Highmore, his French colleagues Louis Lamorier and Louis Bernard Brechillet Jourdain, the German physician Samuel Thomas von Sömmerring, and the Italian anatomist Antonio Scarpa.[35]

The only notable *mechanism* of odor detection was advanced by the French physician and anatomist Hippolyte Cloquet in his 1821 work *Osphrésiologie: Ou, traité des odeurs, du sens et des organes de l'olfaction.*[36] Cloquet recognized the importance of mucous after reading preliminary works on the subject by the German physicians Konrad-Victor Schneider and Johann Friedrich Blumenbach.

Cloquet may have been the first to outline a mechanism of odor detection involving the mucous:

> Once odorous molecules are in the nasal cavities, they spread throughout the area, facilitated by their passage through a narrow opening into a more spacious cavity; according to all laws of hydrodynamics, these conditions should slow their movements and prolong their contact with the olfactory mucosa. They then combine with the mucous, which seems to have physical properties such that the affinity with the odorous molecules is greater than that with the air. The mucous thus separates them from this fluid and traps them on the membrane, where they act on the olfactory nerves, which in turn transmit to the brain the impression received.[37]

Olfactory physiology was mostly uncharted territory until the end of the nineteenth century. Such oversight was not coincidental. It seemed hard to measure and visualize how smells were *processed*. Think about it: How would you determine *where* in the nose odor particles interacted, let alone how? The physiologist Eduard Paulsen wondered about this issue.[38] In 1882, Paulsen came up with an idea. To modern ears, it sounds a little gruesome, yet it was an ingenious experimental setup. Paulsen obtained a corpse head to cut in half (Figure 1.1). He plastered the nasal cavity with small strips of litmus paper before inserting an artificial breathing apparatus. The head-halves then were joined—crude, but effective. Next, the mock respiratory system consisted of a metal tube, serving as the trachea, and a pig's bladder as lungs. Paulsen sprinkled ammonia into the air while sucking the air into the nose. Looking where the litmus paper changed its color, he could trace airflow patterns in the nasal cavity—right up to the epithelium as the locus of odor detection. Zwaardemaker conducted a comparable study shortly after.[39] Instead of a human corpse, Zwaardemaker used the cast of a horse cadaver's head,

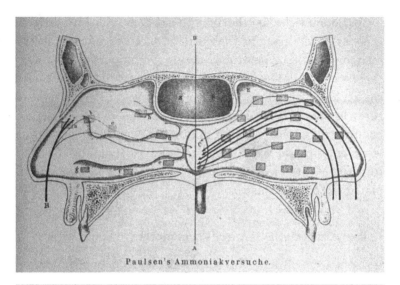

Paulsen's Ammoniakversuche.

Figure 1-1 Cross-section of a human head. In 1882, the scientist Paulsen recorded airflow patterns in the nasal cavity. Using litmus paper (squares on the right side) and an artificial breathing apparatus (extension below on the A-B axis), Paulsen traced the pathway of air sprinkled with ammonia via the discoloration of the litmus strips. Source: H. Zwaardemaker, *Die Physiologie des Geruchs* (Leipzig: Verlag von Wilhelm Engelmann, 1895), 47.

separating the nasal cavity with a glass plate. He placed a candle below the nostrils and pumped in the carbonated air via a breathing apparatus similar to Paulsen's. Zwaardemaker then measured the concentration of carbon emissions from the candle on the glass plate to reconstruct airflow patterns.

Both Paulsen and Zwaardemaker tested for experimental artifacts. Neither the corpse head nor the cast replica contained living tissue. Airflow recordings in pulsating membranes were expected to differ substantially from dead, deflated tissue or rigid inorganic matter. Replications involved several modifications; for example, the experiments used different materials for the inhalation apparatus (rubber tubes instead of metal).

Without insight into the constituents of the olfactory pathway, theories of odor remained speculative. Access to olfactory neuroanatomy hinged on new tissue staining techniques. Chief among them was Camillo Golgi's silver staining method, which allowed the depiction of individual neurons and their differential structures. Ramón y Cajal's detailed observations with Golgi's method continue to be informative to this day (more on Golgi and Cajal's contribution is shared in Chapters 2 and 7). Gradually, the olfactory part of the brain was rendered visible. But what about an explanation for what was seen?

Early Twentieth-Century Psychology

It was far from obvious what smell was. Throughout history, odors had been the invisible essence of things. But what was the essence of smell? What kinds of qualities was smell perception really about? What constituted the psychological interpretation of minuscule molecular features and delicate neural structures? What made a specific number of double bonds or carbon chains turn into mental images of roses or peaches? Without warning, the sense of smell became the subject of psychology.

Early twentieth-century psychology was unprepared for this task. Popular theories, especially psychoanalysis, were occupied with grander generalizations about human nature and society. Smells did not fit that aim. Sigmund Freud, who notoriously associated everything with an expression of sexuality—or the repression thereof—hardly mentioned olfaction. It played a minor role in the analysis of modern humanity. Freud linked the rise of human civilization to the adoption of bipedalism and, regarding the elevation of the nose from the ground, saw a declining importance of smell in human psychology. Besides, Freud reduced any prolonged preoccupation with odors in adult humans as an abnormality and marker of archaic instincts (in addition to a fixation on anal sexuality).[40] He left it at that.

This dismissive attitude was not an exception. When smell did attract attention, it occurred in the context of psychosis, especially female psychosis—such as in the *Treatise on the Nervous Diseases of Women* by the English doctor Thomas Laycock in 1840.[41] Anthropological scholarship strengthened the prejudice that a keen sense of smell was an attribute of primitive cultures and opposed to civilized mankind. But the history of psychology is not limited to Freud and his followers. Other voices existed.

Havelock Ellis, a revolutionary English physician who pioneered scientific interest in human sexuality, was one. His 1905 work "The Sexual Selection in Man: Touch, Smell, Hearing, Vision," in *Studies in the Psychology of Sex,* reversed traditional hierarchy of the senses and their order of dominance even in its title.[42] Ellis dedicated six chapters to smelling. Of course, the intimate connection between sex and smell had always been a mythical jest. The characteristics of the nose, Ellis did not hold back to comment, instigated wild imaginations and rumors throughout its social history: "The Romans believed in the connection between a large nose and a large penis . . . the physiognomists made much of it, and licentious women (like Joanna of Naples) were, as it appears, accustomed to bear it in mind, although disappointment is recorded often to have followed." Beyond myth, Ellis stressed that odor perception in humans was, contrary to popular belief, "exceedingly delicate, though often neglected"—although he likewise pointed at its downgraded sensitivity and purpose in comparison with odor perception in other animals.

Its high variability in associations led Ellis to conclude that smell was "the sense of imagination." This link with imagination was grounded in the suggestive power of scent, "the power of calling up ancient memories with a wider and deeper emotional reverberation."[43] Ellis also highlighted the importance of body odor as a secondary sexual character: "all men and women are odorous." Personal odor communicated identity and familiarity, as "a personal odor resembles a personal touch." Still, Ellis—like his predecessors—found that advanced sensitivities to smell were often linked to abnormal nervous conditions. Few aimed at a theory of odor. The German philosopher and psychologist Max Giessler devised a guide toward a general psychology of smell, *Wegweiser zu einer Psychologie des Geruches*, in 1894.[44] Giessler categorized smell concerning its cognitive and physiological effects. Different odors affected body and mind in different ways, Giessler reasoned. Some odors evoked strong physiological reactions (like coughing, tearing, sneezing, vomiting, or even urinating). Other odors animated specific organ complexes (nerves and muscles) or vegetative systems (respiration, digestion, or reproduction). Smell affecting nerves and muscles often carried an identifying or socializing function. For example, "socializing smells" facilitated the recognition of bonds between individuals, or they familiarized people with their environment and its objects. In comparison, odor with an "identifying" function motivated to direct attention to the source of an odor, a process often building on memory. Meanwhile, smells that, on Giessler's account, interacted with vegetative processes, frequently involved gastric or erotic odors. The category of vegetative odors is wherein Giessler's original contribution lies.

Giessler framed "idealizing odors" as smells with cognitive effects. Idealizing odors involved three forms of cognitive refinement. First, "studious odors," like tobacco smoke, went hand in hand with logical enhancement. Second, odors could result in aesthetic enrichment, given the fact that some odors produced and reproduced abstract mental images. Last, odors could increase ethical behavior either by having a calming or stimulating effect. Each type came in antithetic pairs.

Giessler abstracted odors from their effects without recourse to their material origins (botanic or chemical). But this theory was flawed. It speculated without experimental support. This approach was not unusual for the time; experimental psychology emerged only in the mid-nineteenth century, originating with Wilhelm Wundt and his student Edward B. Titchener. Wundt and Titchener focused on vision, though. Olfaction was of little to no concern in mainstream psychophysics, which aimed at empirical measures of the relationship between physical stimuli and their perceptions.[45] It was a woman who would change this.

The American Eleanor Acheson McCulloch Gamble was the first scientist to look at perceptual dimensions in human smell performance from a proper experimental setting. Gamble was trained by Titchener. In her 1898 dissertation, "The Applicability of Weber's Law to Smell," Gamble tested the responses of several human subjects to odor mixtures, including influential factors like regularity of exposure. Hers was a pioneering approach. At the time it was unclear whether smell would adhere to any psychophysical measure at all. Gamble recognized the predicaments in defining objective differentiae for what appeared to be subjectively perceived distinctions, such as weak or strong odors. Methodological frustration arose from the complex interplay between physiological conditions of the organism (exhaustion, daily variations in sensitivity) and different perceptual effects (odor intensity, odor hedonics, odor quality). She noted:

> (1) Weak smells have vague differences of intensity. For example, vanilla and coumarine soon reach a maximum of intensity which cannot be increased. Greater concentrations simply become unpleasant. (2) Individual differences are more evident for weak smells. (3) The daily variations of sensitivity are more evident for weak smells. (4) Exhaustion has more effect on weak smells. (5) Strong smells hide the weak.[46]

Gamble established systematic measures for experimental comparisons of odor, laying the groundwork for olfactory psychophysics. She determined the chief unit in perceptual discrimination tasks in olfaction, the just noticeable difference (JND). The

JND determines the smallest difference one can perceive in a comparison between stimuli. Gamble's venture into olfaction was part of her broader interest in memory and recognition. Her main concern revolved around the establishment of empirical standards in psychology. Gamble took care not to engage in premature speculation regarding the psychological dimensions of odor. Experimental rigor in psychophysics built on a precise administration of physical stimuli and the testing of their effects. That required new tools in research on smell. Unlike vision and audition, the olfactory stimulus evaded discrete application. Scents can behave promiscuously under varying conditions regarding their environments or interaction with other smells—for example, when mixing with background odors or with previous, lingering fragrances in a room. But Gamble was in luck. One year before her dissertation, Zwaardemaker had introduced a new instrument, the olfactometer (invented 1888, published 1895).

The original olfactometer consisted of a porous porcelain cylinder in form of a long round tube, surrounded by a glass pipe (Figure 1.2, top). Using a pipette, the experimenter filled the space between the cylinder and the pipe with an odorous liquid, before sealing it with cork. This prevented the mixture from interacting with any atmospheric odors in the laboratory. The liquid slowly infused the cylinder. The olfactometer also allowed for adjustments in exposure to different concentrations of the odorous mixture. Zwaardemaker's instrument underwent several modifications. Initially, it had one, and later two, metal nostril pipes. Subsequently, it featured a metal plate between cylinder and nostril pipe to prevent subjects from being influenced by visual cues.

In the absence of a theory of smell, the German psychologist Hans Henning recognized that meaningful assignment of behavioral responses remained improbable. His 1916 landmark habilitation *Der Geruch* offered the first substantial theory of odor backed by experimental research.[47] With little regard for his col-

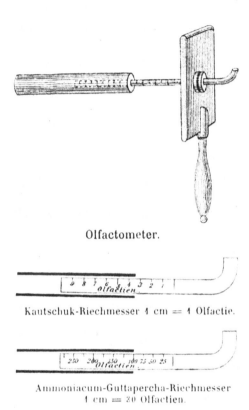

Olfactometer.

Kautschuk-Riechmesser 1 cm = 1 Olfactie.

Ammoniacum-Guttapercha-Riechmesser
1 cm = 30 Olfactien.

Figure 1-2 Olfactory measurement. Top: Zwaardemaker's original design for the olfactometer, invented 1888 and published 1895. Bottom: Scale representing the units of smell measurement (olfactie, the threshold of olfactory stimulation). Source: Top: H. Zwaardemaker, *Die Physiologie des Geruchs* (Leipzig: Verlag von Wilhelm Engelmann, 1895), 85; Bottom: H. Zwaardemaker, *Die Physiologie des Geruchs* (Leipzig: Verlag von Wilhelm Engelmann, 1895), 136.

leagues, he polemically criticized their qualitative terminology, especially in regard to Gamble (his greatest rival). Henning's data collection was remarkable. He tested 451 simple smells and fifty-one mixtures on eighteen subjects, with children and adults of both sexes (recruiting colleagues, their children, and his students). Individuals underwent olfactory training with reference materials for odor classes—for example, violet for floral, lemon for fruity, sulphureted hydrogen for putrid, nutmeg for spicy, frankincense for resinous, and tar for burned. Ambiguous results were replicated with forty-six extra subjects (students).

Henning advanced six primary odors as principal categories. For criteria of odor dimensions, he compared smells with color, sound, and taste. Henning was inspired by the system that Albert Munsell had designed to order color in the first decade of the twentieth century. Munsell's color system was used as an analogy to model three dimensions of odor: the general quality of scent, its intensity, and its clarity or simplicity. Unlike color, however, the blending of smells did not fit the compositional rules for mixing primary colors (for example, green as a combination of yellow and blue). Henning used an analogy with sounds in which odor blends were compared with the "tonal fusion" of chords. To integrate these analogies, Henning reasoned that olfactory space resembled gustatory space (Figure 1.3, bottom). Specifically, smells exhibited a "transitional character" (between corresponding categories) similar to tastes—namely, salty and sweet, salty and sour, salty and bitter. This idea became the smell prism with six primaries: flowery, fruity, rotten, spicy, burnt, and resinous (Figure 1.3, top).

Henning's program was original in its rejection of traditionally linear classification, proposing a three-dimensional representation of odor quality instead. Notwithstanding its popularity, the prism's overelaborate conceptualization evoked criticism. Gamble noted dryly that Henning's idea was suffering from one non-negotiable flaw: "Its very neatness is against it."[48] It was

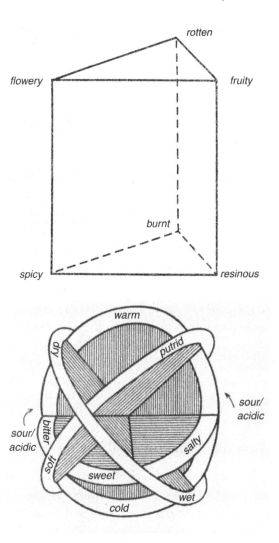

Figure 1-3 Conceptualization of perceptual space in olfaction. Top: smell prism with six primary odors. Bottom: cross-modal sphere integrating smell, taste, and touch. Source: Top: H. Henning, *Der Geruch* (Leipzig: Verlag von Johann Ambrosius Barth, 1916), 94; translation by Barwich; Bottom: H. Henning, *Der Geruch* (Leipzig: Barth, 1916), 26; translation by Barwich.

too simple; the prism did not square with the complexity of mixtures.

Perhaps because of its simplicity, Henning's prism took off like a rocket. It was picked up by subsequent odor studies and classifications, in particular, the 1927 Crocker-Henderson system that reduced these six principal odors to four: fragrant, acid, burnt, and caprylic (fruity-acid).[49] Ralf Bienfang further explored analogies between the senses as a methodological tool. Comparisons with Munsell's color system were especially popular, and Bienfang's "Dimensional Characterisation of Odours," published 1941, detailed the three dimensions of color with odor, specifying the circumference of note (like hue / quality), the axis of clarity (like value / brightness), and the radius of strength (like chroma / saturation).[50]

Henning's methodological contribution that outlasted his prism was his argument for testing both nostrils together. Contemporaries commonly tested smell abilities either with only one nostril open (monorhinic), or first with one nostril open and then, separately, with the other (dichorhinic). Henning rejected this practice as unnatural to human behavior. His principal proposal, the advancement of primary odors, would die before the end of the twentieth century.

The First Half of the Twentieth Century

Despite these first forays, olfaction in psychology and neurophysiology remained a niche subject. Chemistry was the exception and became paradigmatic for research on smell throughout the twentieth century. This is not surprising. The chemical stimulus seemed the only objective and controllable means to measure, quantify, and categorize smell. Nonetheless, chemistry without any model of biology was faulty. One way or another, the sensory system governed the causality of the stimulus: Which structural features of chemicals were responsible for odors?

Many hypotheses emerged at the outset of the twentieth century.[51] Speculations about the structural basis of smell drew on known physical effects. Comparisons with light, sound, and heat resulted in tentative formulations of vibration theories of odor in the nineteenth and early twentieth century. Concrete models and justifications differed widely. Vibration theories included the aesthetic similarity between color and smell, the assumption of a medium like the ether, even the principles of chemistry like Mendeleev's law of periodicity. How did an organism detect such odorous vibrations? One model suggested a correspondence between the vibrations of olfactory cells and odorous molecules. Another proposed that olfactory cells were vibrating as an effect of chemical activity.[52]

The emergence of vibrational theories mirrors the dominance of physics as the benchmark of science at the time. Vibrations encompassed various definitions; they were considered to resemble rays of short wavelengths, Roentgen rays, or light as electromagnetic waves.[53] The first systematic outline toward a vibrational theory of odor originated with Malcolm Dyson in the 1920s and 1930s. Dyson's approach built on the discovery of the Raman effect on light diffraction and photon emission. Robert Wright revived this idea in the 1960s.[54] Still, the biological mechanism supporting such theory remained vague. Hypotheses ranged widely. One idea suggested action at a distance to some medium resonating with the olfactory nerves.[55] Another idea posited mechanical stimulation of olfactory hairs into different vibrations through the weight and momentum of olfactory particles.[56] In the 1990s, this idea attracted attention once more with a nifty quantum physical model by Luca Turin, involving inelastic electron tunneling, which was soon dismissed.[57]

Other proposals combined physics with chemical models. Some speculated that hypothetical receptors could lose energy because of the infrared absorption characteristics of substances.[58] Others suggested that dipole molecules would become neutralized after contact with the membrane.[59] Last, one model purported that vibration frequency was linked to the diameter sizes of pigmentation grains in the olfactory membrane.[60]

Chemical theories of olfaction concerned the interaction of odorous molecules with the epithelium. Explanations oscillated between adsorption (molecules adhere to the membrane surface) and absorption (molecules permeate the membrane surface or are dissolved by a fluid covering it).[61] A less popular hypothesis mentioned the reduction of surface tension.[62] Another addressed water or lipid phases within the membrane or cell itself.[63] An immunological theory of olfaction tested the notion of odor antibodies through tissue injections with insulin.[64] Another looked at analogies of odor stimuli and narcotics.[65] By the 1950s, some scientists started entertaining the idea that a reaction chain of enzymes catalyzed olfactory reactions.[66]

By the mid-twentieth century, there were almost as many olfactory theories as olfactory researchers. What unified these diverse ideas was the central conviction that there must be an intrinsic structural principle determining why a specific molecule carried its particular scent. What that principle was remained a mystery.

Mid-Twentieth Century Onward

Targeted interest in olfaction formed in the mid-twentieth century. At first, advances in olfaction still depended on individuals picking up the topic here and there. The aroma chemist John Amoore, at one of the four U.S. Departments of Agriculture laboratories in Berkeley, was one of these pioneers. The Cornell flavor chemist Terry Acree remembered a chance encounter with him as a student. "This USDA lab was mostly interested in improving the quality of food for civilian and military populations in emergencies of different kinds, including wars and floods and things like that. They spent a lot of time working on preservation. John Amoore was interested in what happened to the flavor of this food as it was frozen, dried, and stored in different ways."

Amoore's interest reached beyond food chemistry. He wanted to understand smell. In the 1960s, Amoore revived the idea of "odor primaries" in light of structural discoveries in chemistry.[67] The biology of the system was an uncharted black box. So Amoore combined two data sets from chemistry and psychophysics to propose five to eight primaries. He devised a way around the lack of biology with an ingenious strategy: by studying anosmics, people who lost their ability to smell. Amoore tested people with specific anosmias, who have an otherwise normal sense of smell but are unable to perceive one or more specific odors. Some people cannot perceive musks, for example. Back then, this was a pioneering approach. Leslie Vosshall, a neuroscientist at Rockefeller University, remarked: "Amoore was a giant in the field to try to wrestle with this problem of how do you go from molecule to percept. His ideas were way before their time. His thinking about specific anosmia perhaps told us something about the mechanisms by which molecules somehow get into the brain."

Amoore hoped to match odor categories with structural classes of chemicals *ex negativo* via their anosmic blueprints. He further speculated about possible receptor sites complementary to primary odors, utilizing the lock-and-key analogy of ligand binding that had become popular with his contemporaries. This model stipulated that ligands would bind to complementarily shaped receptors. The lock-and-key mechanism was initially introduced by 1902 Nobel laureate Emil Fischer in 1894; Linus Pauling suggested that it might also apply to biochemical interactions in olfaction.[68] Next to Pauling, the Scottish chemist Robert Moncrieff worked on a similar structural hypothesis about olfaction in 1949, evaluating the steric (geometrical) properties of odorants (smelly molecules) and reinforcing the increasing explanatory centrality of molecular shape for biochemical research.[69] This marked the beginning of a preliminary theory of odor.

Chemists worked feverishly to find the general rules and details that connected structure to odor, potentially revealing odor classes indicative of receptor types along the way. Another pioneer in structure-odor modeling was the German Günther Ohloff. The chemist Christian Margot at Firmenich remembered working with Ohloff: "He built theories.[70] He was very passionate about research, demanding and always encouraging. An actor of independent, high-quality research." Ohloff was the first to find something close to a structural odor rule, the so-called triaxial rule for ambergris.[71] This rule, published in 1971, stated that odorants smelling of ambergris are defined by the presence of decalin, a bicyclical compound, and that specific atom groups (in three designated positions) must be axial. After its initial success, this rule underwent several modifications. Ohloff's rule faced noteworthy exceptions (like Karanal, a molecule scrumptiously rebutting the above definition of chemical topology). The same fate soon befell other structural odor rules.[72]

Chemists soon acknowledged the striking structural diversity among odorants. Insight into the chemical world of odor grew disproportionally in the second half of the twentieth century, helped by vast technological advances like gas chromatography and the work of many dedicated chemists, including Charles Sell and Paolo Pelosi.[73] This structural diversity thwarted even the most meticulously designed structure-odor rules (SORs).[74] SORs could not capture the code in the nose. Such rules seemed close enough to the causal principles that made molecules smell the way they do. But they were not leading toward the underlying causal basis of smell. Acree nodded: "The point was that studying molecular structures of ligands was not telling us anything about the responses to those ligands in solution in real systems."

In addition to SORs, the lock-and-key model also turned out to be inaccurate. This model's lasting success was that it situated olfaction within the broader research channels in biochemistry.[75] Soon biologists joined the field, among them the physiologist Maxwell Mozell at Syracuse University. Mozell's early theory of olfaction in the 1950s to 1970s and onward embodied the chief research strategy of its time: the search for selective activation via spatial patterning. Inspired by Edgar Adrian's early physiological recordings of spatial activity in the bulb, Mozell's theory compared the nasal epithelium with the function of a chromatograph (the "chromatographic hypothesis").[76] "That's where I got the idea that there may be a spatial relationship in olfaction, of the odorant and the way it's picked up and diffused—like other sensory systems: audition, touch, and even taste to a certain extent." He hypothesized that odorants did not spread over the entire epithelium, but that the epithelium had different zones where odorants interacted. Differences in absorption rates would indicate variations in receptor sensitivity.

Testing frog noses, Mozell studied airflow patterns and found gradations in "sorption" that related to the composition of the chemical stimulus: "Let's suppose that the chromatographic effect has a major effect on what you perceive. I don't know if you read my paper, but I replaced the column of a gas chromatograph with the frog's nose. What I did was to look to the retention of different odorants, all with the regular column for the gas chromatograph. I replaced the column with a frog's nose, and what we found was that they were almost identical." Mozell analyzed differences in molecule sorption, which linked to molecular features such as fat and water solubility. Mozell never got to finish his theory, yet it fundamentally instructed the study of airflow dynamics and flow currents in the nose.

By the 1970s several physiologists entered olfaction. It was uncharted territory, although with slow progress. The growth in biological research initiated a change in the community. This change was social in addition to experimental.

Modern smell science arose as a manageably sized group of exceedingly committed individuals who often found each other at the sidelines of larger meetings, such as the chemical senses group of the United States-based Society for Neuroscience, but also at more specialized meetings such as the Gordon Research Conference (with a thematic focus on the chemical senses of smell and taste every three years). Two meetings specializing on chemosensation had formed by the late 1970s: the European Chemoreception Research Organization (ECRO), which met annually, and the international gathering of the International Symposium on Olfaction and Taste (ISOT), convening every three years. "But none of them really got everybody involved," Mozell recalled. "Because the psychologist went to the psychology meeting, the physiologist went to the physiology meeting . . . there was the Gordon conference that occurred every three years. But you didn't see those people very often. So there was always talk that we should have some yearly annual meeting for olfaction, but nobody ever did much about it."

Change was driven by necessity. In the late 1970s, the United States-based National Science Foundation (NSF) was cutting back funding, and massively so. Research on the visual system hardly needed to fear for its financials, as it was a hot topic in the burgeoning field of neuroscience. Funding for olfaction was neither stable nor secure. It already existed on mere life support. The North American community required organization to attract money.

"I was doing a stint at the National Science Foundation, NSF, and I was working with a guy who kept chiding me: 'You guys in olfaction haven't learned as much compared to the vision and audition people.' He was getting under my skin a bit. But he was right because at that time we knew very little. We knew that molecules brought odors, but we didn't know much more than that. So, we [in the field] kept talking when we got together at the physiology meetings. We ought to do something. We were to do something. But then NIH [the United States-based National Institutes of Health] and NSF both did something." Funding policy changed. Olfaction needed support. "The idea was that we should also have a society that can talk to NIH and NSF and represent the research going on in the United States."

The Association for Chemoreception Sciences (AChemS), was born. "There was no society that brought them all together, until AChemS," Acree recalled, "which was a brilliant thing that happened in my lifetime—it allows these fields to develop in parallel within the structure of the society." It was a frequent platform to compare experiments, develop ideas. In short, AChemS, in addition to ECRO, brought some form of connection and continuity into the community, accelerating research. Mozell remarked that, in comparison, the community looks small: "AChemS has about five hundred members. Vision goes into millions." Today, AChemS is an institution. In 2018, the meeting celebrated its fortieth anniversary. "So, if it wasn't for the fact that NIH and NSF decided that more money should go to vision and audition rather than the chemical senses, we may never have become an organization!" Mozell laughed.

By the late 1970s and early 1980s, the field had finally come together. Olfaction was no longer a playground solely for chemists but opened its doors to biology. In the 1980s, systematic inquiry into the molecular basis of the olfactory system began. Several studies suggested that odor detection relied on the same molecular pathway as the detection of stimuli in other senses. External information from the stimulus (light photons, sound waves, or airborne molecules) gets transformed into electrical signals in the sensory nerves. Responsible for this transformation is a so-called second messenger pathway: a cascade of biochemical reactions that, in its various molecular constituents, was stepwise revealed by several biochemists, geneticists, and neurobiologists. "There had been this huge move towards molecular biology in olfaction," neuroscientist Stuart Firestein at Columbia recalled. "I especially think of Randy Reed at Hopkins, Han Breer in Germany, Doron Lancet in Israel, Gabriel Ronnett, several other people like that, who were doing hardcore cellular, molecular biology in olfaction and who really figured out the entire transduction system. This is before we had the receptors; it was the somewhat more tractable question at the time."

These findings discredited the belief that the sense of smell represented something like the odd one out: "Early on, when I first came into the field," remembered Gordon Shepherd, Yale neuroscientist and previous mentor of Firestein, "olfaction was way off to the side. Nobody knew anything. As they started doing it, everything seemed different. Because you couldn't imagine what the smell stimulus was like, you had troubles controlling it . . . You didn't know *what* was happening in the brain. People thought olfaction was special, that it was different. As the groundbreaking work was done in vision, it was assumed that vision was a mainstream sense, and that olfaction was a sort of side kind of sense."

With growing insight into a common mechanism of molecular communication, it stood to reason that smell could be subject to the same general principles governing other sensory processes. This revelation started to elevate olfaction from a niche closer to the margins of mainstream science. It further initiated a sudden race. A key but missing piece of the puzzle involved the membrane receptors that started this molecular gateway.

It would take another decade before these efforts would truly pay off. In 1991, with the long-anticipated discovery of the olfactory receptors by Linda Buck and Richard Axel, the field of olfaction finally hit its jackpot. And what a jackpot it turned out to be. The discovery of the olfactory receptors, almost overnight, would catapult olfaction from obscurity into the spotlight.

Whatever alternate histories could have played out, the modern science of smell is radically divided into two eras: the time before and after the receptor discovery. One may even ask whether this discovery arrived just in time. The neurobiologist Randall Reed at Johns Hopkins University speculated: "The greatest danger I thought to the field was that we would have gone another decade without finding receptors and people gave up. If you think about what happened if Linda just said: 'I give up.'" But she did not.

This is where the story of present-day olfaction begins.

CHAPTER 2

Modern Olfaction

AT THE CROSSROADS

The breakthrough defining the modern era of olfaction occurred in 1991. Linda Buck and Richard Axel discovered what would eventually be identified as the largest multigene family in the mammalian genome.[1] This discovery did not come easily. Buck searched three years for the olfactory receptors (ORs). Because of the significant diversity of odor chemicals, she thought there must be a sizeable receptor family, although it was inconceivable just how sizeable this family would turn out to be. Buck initially found a number of different receptors. None belonged to the elusive olfactory family. No other lab was succeeding either. Still, three years without publishable results is a long period, especially for a scientist early in her career.

Buck's boldness, therefore, should not be forgotten in hindsight. Stuart Firestein, a friend of Buck, highlighted the importance of Buck's persistence in a laudation at the Harvey Society's Lecture Series:

> I was working in olfaction for some years before anyone, except Richard, knew of Linda Buck. . . . I appreciate Linda because she is to me the portrait of courage in science. I use her as an example with students. She went after a result that

had no intermediate, no publishable alternative. Now Richard, who is here tonight and paid the bills, supported the work and had the vision to see how important this was, but it's true to say that if someone else in some other lab had found the ORs he would not have disappeared into obscurity. But for Linda, a postdoc of some unmentionable number of years, those were the stakes she was playing for. She was betting the house—and perhaps her scientific career. In the current environment that emphasizes translational research, generating licensing fees and doing something "useful," it is harder to come by such examples of scientific courage. Linda reminds us that bravery works.[2]

Richard Axel remembered the moment Buck stepped into his office with her findings: "She devised this very clever scheme, and she got it. When she showed me the results, I was silent for a while because the whole thing began to unfold in my head."

A quantitative expression for the importance of this discovery is the Science Citation Index. In the thirty years before Buck and Axel's breakthrough, 295 research articles used the keywords "odor" and "odor receptor"; in the five years after their publication, the number was up to 406; over the past twenty-three years the number has grown to 4,037. Their original publication was further selected for a series of annotated research papers in *Cell*, celebrating fundamental breakthroughs in biology over the past forty years.[3] Meanwhile, Buck and Axel received the 2004 Nobel Prize in Physiology or Medicine.[4] Olfaction, a minor footnote in the history of science, was catapulted into mainstream research.

The Nobel Nose

What was so extraordinary about the olfactory receptors? How did this discovery lay the foundation for current olfactory neuroscience? Three reasons mark the significance of these receptors: their sheer size, the method of their discovery, and their experimental role of providing systematic access to the olfactory brain.

The sheer size of the olfactory receptor family links to the larger protein family to which it belongs: so-called G-protein coupled receptors (GPCRs). GPCRs are a superfamily of transmembrane proteins involved in a range of fundamental biological processes like vision, the regulation of immune responses, and the detection of neurotransmitters. Today we know that this protein gene family occupies up to about 10 percent of the mammalian genome. But the significance of this gene family began to emerge as early as the late 1980s, when the molecular biologist Robert Lefkowitz at Duke University reported that the adrenalin receptors and rhodopsin share a significant range of highly specific structural motifs, and might constitute a much larger receptor family.[5] Many scientists had hoped that olfactory receptors, as GPCR candidates, could yield a few interesting genetic findings. These hopes were not fulfilled—they were outdone.

Membership in this family opened up research on smell to mainstream science. The structural and functional features of the olfactory receptors proved an excellent paradigm for research on GPCRs. Exceeding all estimations, the size of the olfactory family turned out to include about one thousand members in mice and four hundred in humans. Putting this number into perspective, the biggest GPCR receptor family before the odor receptors was that of serotonin—with a slightly less dramatic number of fewer than a dozen members known at the time (today their number is fifteen). Another intriguing revelation about the genetic makeup of these newly found receptors was that they share several key amino acid motifs with other members of the GPCR superfamily. Still, odor receptors also share additional motifs with each other and appear to be highly variable among the members of their own group. In other words, the olfactory receptors mirrored both the functionally and structurally most salient properties of GPCRs, only on a smaller scale, and they comprised a separate family. Considering that about 50 percent of current research on drug design targets GPCRs, the implications of cracking the molecular code of the olfactory receptors reach farther than the nose. On a molecular level, odor detection did not signify the odd one out. It was a prime model.[6]

Genetic interest in the olfactory receptors was heightened also by the discovery method. Olfaction became an accredited part of mainstream neurobiology because it provided proof for the successful application of a principal genetic technique, including an extension of its utility.

Buck's stroke of genius was her experimental design. It built on an unprecedented application of polymerase chain reaction (PCR). PCR is a method based on the natural process of DNA replication, which involves an enzyme (polymerase) duplicating DNA strands targeted by primer pairs. Primers are short sequences of nucleotides that bind to specific genome sequences in a complementary fashion. This procedure can be replicated exponentially through repeated reaction cycles, producing vast amounts of particular gene strands. The distinct advantage in the invention of this method is that it solved the issue of scarcity of genetic material.

Few technologies have been as revolutionary as the invention of PCR. This invention by Kary Mullis, recipient of the 1993 Nobel Prize in Chemistry, was "virtually dividing biology into the two epochs before P. C. R. and after P. C. R."[7] PCR was a relatively new tool during Buck's postdoctoral years. "When the PCR papers came out, I was thrilled," Buck recalled. "I thought that PCR would open up the door to many things. Like a miracle drug for molecular biologists to use! Just think of what the first microscope allowed people to do; they could look, they could see things. And to me, it is all about being able to see things!"

The early days of any new technology face manifold challenges in fitting the materials to the method and determining a method's range. At the time, PCR did not look like the most obvious tool to discover an unknown gene family. This method piggybacks on a natural copy-and-paste mechanism and amplifies known genetic materials, allowing for the production of sufficient materials for large-scale genetic research. Essentially, *the* precondition of PCR is that the genome sequences one intends to amplify must already be established, at least in parts—but none of the genome sequences for the olfactory receptors were known at the time.

Buck combined two modifications of PCR. Like a fishing net, she used a combination of variable genetic patterns (called degenerate primers) to tag and copy a range of diverse but similar-enough genetic sequences. Primers in PCR are called degenerate when some positions of their sequences have more than one possible base: "for example, in the primer GG(CG)A(CTG)A the third position is C or G and the fifth is C, T or G."[8] The degeneracy of a primer describes the number of its unique sequence combinations (six in the example). Degenerate primers, therefore, are less specific and allow for amplifications of related yet heterogeneous genetic sequences.

"For the degenerate primers, I collected all those sequences of the known [GPCRs], which was a very limited number, and aligned them by hand. I then designed degenerate primers that give you combinations and have the capability of amplifying any of those GPCRs." She went one step further: "When it came to the general primers, I thought, 'Maybe they're GPCRs, but maybe they are some other kind of receptor, maybe the nuclear type receptors.' So I actually designed the general primers not only for GPCRs but also for the nuclear receptor family."

Buck's trick was not to look for the known motifs common to all GPCRs. Her combinatorial mosaic searched for partial, overlapping resemblances across different GPCRs. How would you know whether you found the right genes? Buck's second brilliant move was to use RNA instead of DNA. This choice resulted in different molar weights (concentrations) of the obtained genetic materials and allowed her to pick out the heaviest bits! The use of degenerate primers in PCR soon became part of the standard repertoire of genetics—for example, in cross-species genetic comparisons.

The identification of the receptor genes finally handed the olfactory scientists the key to the olfactory brain. The discovery brought to light a very peculiar feature of the gene expression in the olfactory system, one that promised direct access to its neural machinery. Every sensory nerve in the nasal epithelium expresses (meaning "makes appear in the cell") one receptor gene. Consequently, if an experimenter was able to trace the activation signal of an individual sensory nerve, she can see directly how and where a receptor communicates its signal to the brain.

By the end of the twentieth century, the necessary pieces to decode the nose were on the table. Over the next two decades, several laboratories entered into a fierce competition to crack the olfactory code. Many scientists in the 1990s and 2000s believed that the olfactory brain would reveal its inner workings in virtually no time. The prevailing assumption was that the olfactory system, like any other sensory system, uses neural space in a distinct topographic fashion to represent its stimuli. The open question was how. This question had already been addressed in research on the visual system. There were plenty of reasons to think that the olfactory brain could be modeled analogously to the visual system.

The Paradigm of Vision and Functional Localization

It is impossible not to be enamored of the visual system. Think about it for a minute: What you ordinarily see is not what the cells in the retina "look at." Vision begins with single photons. How does the detection of photons result in complex visual images such as human faces?

To appreciate how the visual system works is to recognize the central principle that the cells in our visual system are choosy. Just any input does not excite them. They respond selectively to particular features. Because of this choosiness, it is possible to single cells out into clusters and infer just how these clusters build on each other successively in the sophisticated machinery of sensory feature extraction. How this works is quite remarkable.

Vision begins when energy patterns in the form of photons hit the receptors in the retina (which for now we imagine as acting like a two-dimensional plate at the back of our eyes). Their signal gets carried through the optic nerve, a bundle of nerve fibers called axons, to the thalamus. If you cut bilaterally through a vertebrate's head, the thalamus is located right at the core of the brain. Similar to a router, the thalamus works like a relay station by collecting the input from various sources, sensory as well as motor information, and transmitting it to the appropriate areas in the cerebrum, the sizeable surrounding part of the upper brain. Here, the information from the retina passes through the visual part of the thalamus, the lateral geniculate body. This information is sent next to the back of the skull, where the visual cortex lies. The visual cortex consists of several different areas, denoting functional subdivisions. The focal point to which the main signals from the thalamus project is the primary visual cortex, known as the area V1 or striate cortex. From there, signal projections disperse to several higher-level areas in the visual cortex (associated with specialized functions, such as self-motion, direction, or color processing).[9]

Research on vision is in full swing and renders this sketch of the visual pathway highly simplistic. The visual neuroscientist Aina Puce at Indiana University Bloomington added that vision is not explained by a shallow retina-thalamus-V1 model: "Of course there is the alternate pathway that runs to pulvinar and superior colliculus and onto extrastriate cortex, bypassing V1. Its existence allows the phenomenon of blindsight to manifest!"

The fundamental principle of the visual pathway is its representational setup. The visual cortex operates on what is called retinotopic maps. These maps denote the cortical areas matching specific regions in the retina: When we are working in the cortex, these maps allow us to locate exactly from where in the retina the signal is coming. Vice versa, if we are looking at the retina, we will see to which spot in the cortex the signal is projected. A helpful illustration of this principle is the fovea, the focal point in the retina where our sight is most acute due to the highest concentration of retinal cones (color-sensitive receptors that respond to high-intensity light sources). Signals from the fovea lead to a discrete area in V1. Having a point of reference, we can then map the input from the surrounding cells in the retina, which are responsible for our peripheral vision, to the rest of the primary cortex. This organization provides a neural mirror of retinal input. Since this idea encapsulates the core of sensory neuroscience, further instructing modern olfaction, we will discuss both its historical appearance and technical details.

The enigma that occupied many brain scientists over the first half of the twentieth century was how information from the external light source gets encoded throughout the various stages of the visual pathway. The sensory system delivers visual data in the form of spatial and temporal patterns that can be recorded through the firing rates of neurons. The question was how the system differentiates and integrates these patterns into a unified visual image.

A revolutionary series of experiments provided a chief point of entry in solving this puzzle.[10] During the late 1950s, two postdoctoral researchers at Johns Hopkins University, David Hubel and Torsten Wiesel, recorded electrical signals from individual cells in the visual cortex of a cat. They inserted a microelectrode in the V1 region at the back of the cat's head and recorded its responses to several light stimuli flashed on a screen. As is customary with cats, not much happened at first. Neither the anesthetized animal nor its cortical cells were reasonably interested in whatever images were shown. This might have been the end of the experiment if not for one of those fortuitous accidents characteristic of celebrated events in the history of science. As it happens, Hubel and Wiesel administered the stimulus with transparent glass slides depicting circular black and white spots via an overhead projector. The story goes that one day, after many hours of fruitless examination, they had started packing things up when, suddenly, the microelectrode in the cat's head produced a sound of cell activity that resembled the rapid firing of a machine gun.

It took several hours to find the reasons behind this phenomenon. The glass slide they were using was dirty at its edges. When it slid off the projector swiftly, it was producing the contrast of a thin line—not a spot—on the screen. With further tests, Hubel and Wiesel noticed another phenomenon. Not only were these striate cells responding to lines, they also seemed to prefer lines in specific orientations. Moreover, and this was key, cells expressing an inclination for such orientation preferences tended to form aggregates.

Whatever you see is not a passively received mirror image of the world. Visual imagery, these findings now revealed, presents a construction by your sensory system. Hubel and Wiesel published their breakthrough in a series of three papers that, in 1981, was crowned with the Nobel Prize in Physiology or Medicine.

The impact of Hubel and Wiesel on the gradually maturing field of neuroscience was twofold. Most importantly, it decisively settled the ongoing battle between two general theories of the brain.[11] One theory assumed that any mental processing was distributed more or less equally over the entire brain, or involved at least most of its areas. This idea is known as the "field theory" and the "equipotential theory of the brain." It dominated the nineteenth century, and some of its prominent advocates were the French physiologist Jean Pierre Flourens and the Swiss anatomist Albrecht von Haller. Despite growing objections, this theory continued to influence research on the neural and psychological mechanisms of the brain over the first half of the twentieth century—for example, in the works of the American psychologist and behaviorist Karl Lashley.

The competing theory was functional localization, which compartmentalized the brain into distinct anatomical domains corresponding to physiological functions and mental faculties. One of its earliest proponents was the Swedish scientist Emanuel Swedenborg in the eighteenth century. The idea of localization became widely known in the form of phrenology and the passionate advocacy of the German neuroanatomist Franz Joseph Gall in the nineteenth century. Skepticism prevailed at first. Subsequent observations, however, chiefly from lesion experiments, supported the underlying hypothesis that some sensory and cognitive faculties were conditional on the activity of specific brain regions. This assumption received further credit through experiments by prominent figures like the French physician Paul Broca and the Scottish neurologist David Ferrier.

The turning point came at the dawn of the new century with the Golgi method, named after its inventor, the Italian physician Camillo Golgi. This staining procedure elegantly demonstrated that the brain was not a homologous substance but consisted of a tangled, intricate net of differentiated nerve cells and layers. (This was not without irony, of course, since Golgi had feverishly argued for the opposite. He believed the whole brain was a syncytium, a continuous mass of tissue that shares a common cytoplasm. His opposition to Ramón y Cajal's neuron doctrine was legendary.) The visualizations obtained with Golgi's method provided indispensable insights into the anatomical differentiation of the brain. Hubel and Wiesel's findings gave necessary evidence for a precise functional differentiation of its distinct cell layers.

The second implication of Hubel and Wiesel's findings even exceeded the impact of the first. Their studies opened up a brand-new way of modeling the neural basis of sensory processing, one in tune with the Zeitgeist. A central concept gaining enormous traction in the biological sciences at the time of Hubel and Wiesel's work was the notion of information. Several lines of research started working with the premise that biological systems operated just like information coding machines, especially in genetics, but also in research on the sensory systems. The significance of thinking about information in biological terms was underscored by the 1954 discovery of the DNA double helix by James Watson and Francis Crick (and Rosalind Franklin to a certain extent). The notion also took off in cybernetic studies of biological systems, as championed by Norbert Wiener, Walter Pitts, and others.[12] In this context, the quest now became to find out what information was communicated in sensory systems. What were the principles of the coding system by which this information was processed from the primary sensory organs, such as the eyes, to the brain?

In the early 1950s, Stephen Kuffler, in whose lab at Johns Hopkins Hubel and Wiesel were working, demonstrated that the first step in visual information coding was taking place already in the cells of the retina.[13] This finding was not trivial because it meant that the eyes did not merely record but actively filtered and organized light patterns. Kuffler's conclusions were in agreement with other studies of his contemporaries—in particular, with Jerome Lettvin's famous 1959 publication with the congenial title "What the Frog's Eye Tells the Frog's Brain."[14] The key Kuffler was handing down to Hubel and Wiesel was the concept of a receptive field. A receptive field signifies the range of the light stimulus, and its position in space, to which the retinal cells respond.

Kuffler's research showed that the retinal cells were constructing the visual signal in a somewhat circular fashion through their arrangement into what became termed "center-surround cells." Some cells respond to light energy with increased activity. These excitatory cells are surrounded by further cells displaying inhibitory behavior when stimulated. Accordingly, such circular formations are labeled "on center." The other group of cells, labeled as "off center," is built conversely. Here we have an inhibitory cell in the middle with excitatory cells forming its surroundings. In a sense, what the cells in your eye depict is a world of light and dark spots. Naturally, this is not the image we come to see. How do we get from such a snowstorm of black and white dots to our perception of clearly delineated objects?

This was the enigma Hubel and Wiesel tried to solve with their studies of the cat cortex. What they found in the cortical cells of the cat was a highly specialized procedure of hierarchical integration. The higher the level of sensory processing, the more selectively the cells responded. Throughout several mediations of synaptic transference, the receptive field of black and white spots in the retina transformed into lines in the primary visual cortex. Similarly, these edges were further assembled into curves, and so forth. By moving from the primary cortex to the higher cortical areas in the visual system, we see a transition from so-called simple cells (orientation sensitive) to complex cells (movement sensitive), up to hypercomplex or end-stopped cells (length sensitive). The increasing input selectivity of cells mirrors developing image complexity.

This advancement substantiated the hypothesis that the brain generates its content from specialized aggregates of cell groups, not from the input of all individual neurons uniformly. The emerging understanding of the brain was that it operated by successively organized and hierarchical stages of neural clusters. The realization of such an orderly stepwise integration of information processing provided the basis for the emerging computational conceptualization of wiring schemes, prompting questions about what was computed by the brain and how.

The introduction of a computational understanding into the study of neural signaling was of major impact for the young discipline of neuroscience. It allowed for the dissection of sensory processing into various stages that, at least in principle, were analyzable more or less separately of each other. It encouraged a theoretical perspective on the individual steps in perceptual processing that was independent of the limited and contingent insights available from the neural hardware alone.

The Bauplan of Neural Topography

The triumph of the visual system as the paradigm for sensory neuroscience lodged in its neural architecture. Resonating with its methodical appearance of computational processing was the meticulous organization of its physical implementation. The entire visual pathway operated by systematically connected groups or populations of neurons forming neural maps, and underlying these connections was a robust topological principle. The principal notion driving topological interpretations of neural specialization in the cortex was the concept of the column.[15]

Cortical columns refer to vertical collections or bands of cells that respond to the same stimulus properties—two kinds of such cell specialization command the visual system: ocular dominance and orientation columns. Like slices of bread, ocular dominance columns preserve segregation of the signals from the left and right eye, starting with their retinal origins and traveling through alternating layers in the lateral geniculate body (thalamus) up to the cortex. Consequently, when you insert a microelectrode into the cortex, it can be determined from which eye it gets its input. When moving the microelectrode horizontally over the cortex surface, re-cordings vary sequentially and in constant intervals between the left and the right eye, in a pattern not too dissimilar from that of a chessboard. By contrast, when the microelectrode is inserted per-pendicular to the surface, recording continues from the same eye, either left or right.

The other kind of columnar formation describes orientation-specific cells. As we saw earlier, cortical cells are highly selective to the light contrast of illuminated lines. These cells further form regular aggregations with striking consistency, such that "every time the electrode advances 0.05 millimeter (50 micrometers), on the average the preferred orientation shifts about 10 degrees clock-wise or counterclockwise."[16] Nearby anatomical positions in cells, therefore, represent a kind of neighborhood in input space.

That idea of anatomically discrete functional units emerged from fundamental discoveries in the somatosensory cortex. Neu-rosurgical procedures revealed locally differentiated areas in the brain. For example, electrical stimulation of specific areas in the sensorimotor cortex can cause a tingling sensation in the foot, whereas stimulations in other designated areas impede arm movement. Pioneering here was the research by Wilder Penfield, neurosurgeon at McGill. In the 1930s, Penfield translated the ar-rangement of discrete functional areas on the brain surface into the idealized human shape of a cortical homunculus. This idea successfully integrated the principles of neurophysiology with the practice of neurosurgery.

During the 1950s, Vernon Mountcastle at Johns Hopkins un-dertook a detailed neurophysiological examination of receptive fields in the somatosensory cortex. Mountcastle championed the idea of a columnar organization of the somatosensory cortex. These columns specialized in the representation of distinct stim-ulus features, such as touch, pressure, or joint position. It was not a readily embraced idea. His proposal was vigorously contested at first—so vigorously that even his collaborators asked to be removed from the original publication in 1957.[17] Mountcastle's observations provoked Hubel and Wiesel's imagination. Their findings in the cat cortex legitimized talk of cortical columns. Before long, similar observations of cortical columns arose in research on the auditory system. To be sure, postulates about columns in the primary auditory cortex had been posed as early as 1929 by the Austrian neurologist Constantin Freiherr von Economo. But physiological proof followed only in 1965 and the following years.[18] Cortical columns undoubtedly catalyzed research toward a unifying model for neural processing, though they remain disputed in their significance.[19]

The growth of neuroscience as a mature field went hand in hand with the emergence of topological modeling. This development accelerated with technological breakthroughs in optical imaging, such as functional magnetic resonance imaging (fMRI) in the 1980s—with its cognitive appeal of spatially processed data—and two-photon imaging of calcium signals in the 2000s. Three concepts were constitutive of this burgeoning paradigm.

The first concept is *information,* the computational organization of sensory signals. By way of illustration, consider visual signals that are transformed into neural messages by the receptive fields of cells, denoting the input range a cell imparts on other cells. Central to a sensory system's formation of receptive fields are the two processes of excitation and inhibition. One role of these complementary cellular mechanisms is the sharpening of receptive fields and generation of cortical oscillations (relevant to functional network states; Chapter 8).[20]

This understanding of information inevitably builds on another central concept, *computation,* the process by which these neural signals become organized into information patterns. What sensory information entails, or what its function is, cannot be inferred without a look at the underlying arrangement of neuronal connectivity.

A quintessential premise specifically for topographic representation is *localization,* modular organization into more or less discrete functional units. We saw how the visual system aggregates cells that respond selectively to similar stimulus features. The most salient manifestation of topological connections between specialized aggregates is retinotopic maps, the areas in the primary visual cortex resonating with positions in the retina.

Topographical projections in both sensorimotor and sensory cortices supported the idea that the brain operated by general principles. The hierarchical organization of the visual system, which lent itself for computational theorizing, especially cemented neural topography as paradigmatic.

From Nasal Nerves to the Olfactory Brain

How did neural mapping in vision become applied to the olfactory system? Parallels to the visual system were already successfully conceived in the receptor discovery. The discovery of a second messenger mechanism and the receptors in the olfactory system drew considerably on advanced insights into signal transduction in vision and the study of visual rhodopsin as cell surface receptors. Further progress from analogous modeling was expected—it was also expected that the receptor discovery would quickly lead to something like a "code in the nose." Such a code was deemed vital to examine the neural correlates of smell. So the first step was to determine the receptive fields of the newly found family of transmembrane proteins. Scientists believed these fields would explain what features were extracted, collected, and topologically represented.

Optimism dampened, however. The complexity of olfaction's molecular machinery exceeded all expectations. Not only was the sheer size of the receptor family beyond all previously known measures but, as Buck and her postdoc, Bettina Malnic, reported in 1999, these receptors also operated by combinatorial coding.[21] Olfactory receptors do not respond exclusively to a specific ligand or uniform group of ligands, such as musk molecules with a ring structure. One receptor can detect a range of diverse molecules by various features and, vice versa, one molecule binds to multiple receptors with different atom groups. The estimate of possible stimulus interactions skyrocketed. Besides, this realization finalized the unlikelihood of "primary odors."[22] The concept was reminiscent of that of primary colors in vision, in which a small subset of colors accounts for the qualities and range of their mixtures. But the combinatorial coding of odors proved the determination of a comprehensible number of receptor primaries futile.

Combinatorics was only one challenge. Another complication was the delicacy of the olfactory receptors—in particular, their expression behavior in cells. An often-overlooked detail in the history of Buck and Axel's breakthrough illustrates this point. Technically, in 1991, Buck and Axel never demonstrated that the multigene family they found really expressed the olfactory receptors. In titling their article "A Novel Multigene Family May Encode Odorant Receptors: A Molecular Basis for Odor Recognition," Buck and Axel were not being modest; use of the modal modifier "may" was deliberate. Although it seemed evident that the genes found by Buck and Axel should be those of the olfactory family, it was not conclusively confirmed. The standard method to affirm the identity of this gene family would have been through heterologous expression. (In this case, heterologous expression involves injecting the genes into nonolfactory cells like those of yeast to make them express the receptors and then testing the cell's responsiveness to a range of odorants in what would have been otherwise unresponsive cells.)

This simple strategy turned out to be a Herculean task. Olfactory receptor genes put up considerable resistance to appearing in any other cell than those of the olfactory sensory nerves. It took a decade and a half to achieve this. Meanwhile, in 1998, Stuart Firestein and his former graduate student Haiqing Zhao found a clever way around this problem.[23] Firestein recalls Zhao's eureka moment: "One day, he ran into my office and said: 'I know which cells we can use to express the olfactory genes. The olfactory neurons!'" Firestein quickly noted the predicament with this idea—namely, that "this had been the very problem all along. Mind you, the olfactory sensory neurons were the only cells expressing these receptor genes." Since the mouse epithelium has about one thousand of these genes, how was one to tell—unambiguously—that an isolated gene expressing a receptor responded specifically to an odorant? In other words, it seemed impossible to know whether the isolated X specifically reacted to Y and not something else.

Zhao and Firestein turned the problem into its solution. Because only olfactory sensory neurons expressed odor receptors, they decided to use olfactory sensory neurons and amplify the expression of one receptor gene to determine its binding range. They infected a rat's epithelium with a virus that carried an isolated receptor gene. Such an infection led to an overexpression of one specific receptor in the epithelium, increasing its appearance from about 1 to 30 percent. Consequently, whatever ligand excited this receptor (known as the rat I7 olfactory receptor today) resulted in a massively disproportionate response, making it possible to determine its binding range.[24] This experiment was a proof of principle. It was too laborious and time-consuming a procedure to apply to more than one thousand genes and hundreds of thousands of ligands.

Receptor expression continues to be tricky. Hiroaki Matsunami at Duke University, a former postdoc with Buck, achieved the first heterologous expression of odor receptors in 2011.[25] Matsunami started focusing on odor receptor deorphanization (the determination of odorants to which a receptor responds). Meanwhile, in 2014, Joel Mainland, at the Monell Chemical Senses Center, refined Matsunami's method to do the first human olfactory receptors expression.[26] Expectations about decoding smell were lowered, but not blocked.

Soon another realization struck. Receptor coding in the nasal epithelium did not compare to the process of the retina—to start with, there were no center-surround cells. Also unclear was what an "odotopic map," analogous to the retinotopic structure, would look like. Ultimately, the assumption of an odotopic map in the epithelium did not materialize. At first, in 1993, Buck and Kerry Ressler did find a rough division of the epithelium into separate zones of gene expression.[27] Unlike center-surround cells in vision, however, these zones did not lead to spatially discrete receptor patterns.

Perhaps the map formed at a later stage. The obvious place was the olfactory bulb, the next step in the olfactory pathway. Already Cajal had made that suggestion: "A careful comparison of the structure, position, and connections of the said [olfactory] centers with those of the corresponding visual, tactile and acoustic systems allow us to recognize that *the first or olfactory bulb is homologous to the retina* (not the whole retina but to the inner plexiform layer, ganglion cell layer, and [optic nerve fiber layer]), to the ventral and lateral [cochlear nuclei] in the medulla, and to the [dorsal column nuclei] of this same center."[28]

Let us look at the eye again to understand Cajal's comparison. Unlike the olfactory pathway, visual information is mediated by numerous synaptic connections and multiple layers of neurons between various cell types before it reaches the cortex. In the retina alone, we encounter three layers of sensory cell neurons, each with a specific function. In particular, we find two types of receptor cells and four types of retinal neurons: the bipolar, ganglion, horizontal, and amacrine cells. Starting with receptor cells situated in the first layer at the back of the retina, we distinguish two main types: rods and cones. Rods are slender formations that all carry the same pigment and react to dimmer, as in less intense, light sources. They are responsible for night vision. By comparison, cones are notably thicker and usually come in three types of pigment that resonate with intense light and facilitate color vision. Following this first layer of specialized receptor cells is a second layer of bipolar and horizontal cells. Here, information from several receptor cells gets collected by the horizontal cells and conveyed to the elongated bipolar cells, which can also receive input directly from the retinal cells. In a third layer, the spherical retinal ganglion cells continue to collect this information from the bipolar cells before it is sent out of the retina through the optical fiber. Similar to the horizontal cells we also have a layer of amacrine cells that partially mediate the information between the bipolar and ganglion cells. The purpose of such multilayered processing is to sharpen the resolution of the receptive field, with a center-surround formation of signal transmission notably being maintained throughout. Walking through such a multispecies forest of interconnected cells makes one appreciate the two-synapse simplicity in the olfactory system a bit more.

The anatomy of the olfactory bulb shows a compelling similarity with the visual system. As does the retina, the bulb operates via different cell types distributed across different layers. Roughly speaking, it has the following structure. Olfactory sensory nerves from the epithelium converge in the layer of so-called glomeruli, spherical neural structures in the first (outer) layer of the bulb. The glomeruli are innervated by two kinds of cells, the mitral cells (titled after their resemblance to a bishop's cap) and their smaller counterparts, the tufted cells. (Innervated means these cells provide the nerves that further project the signal to the olfactory cortex, rendering the first synaptic interface in the olfactory pathway.) Similar to retinal receptors, these glomeruli are interconnected horizontally, allowing for lateral inhibition of neighboring cells (that is, the cells can reduce the excitation of their neighbors). From this view, the composition of the olfactory bulb seemed to resemble the retinal makeup both anatomically and functionally. Proving this hypothesis at first seemed successful: Robert Vassar and Axel soon reported recording spatially discrete and topological patterns of stimulus activation in the bulb.[29]

This bulbar map was not precisely startling news to one man in particular. Gordon Shepherd had advocated for a topographic organization of the olfactory bulb before the receptor discovery. "I was convinced when it became clear that there was lateral inhibition in the olfactory bulb, and that we had the same electrophysiological properties as have been seen in the motor neurons. Eventually, we found the creational patterns representing odors, basic mainstream properties that have been shown in the somatosensory system, in the visual system, and the motor system, and so forth. I think it had to be true also of smell." Shepherd found that "in comparison with the visual pathway, which starts in the retina and processes through the thalamus to the visual cortex, it is as if the olfactory equivalents of all these structures are compressed into just the olfactory bulb."[30]

At that time, however, molecular biologists and physiologists were not necessarily participating in the same discussions or looking at the same literature. So the broader attention that Buck and Axel's gene discovery received did not translate immediately to Shepherd's earlier theories and experiments on the olfactory system and, especially, the bulb. Chapter 1 showed that previous to, and throughout, the 1990s, there was no cohesive community working on the biology of olfaction; rather, it was a cross-disciplinary conglomerate with a variety of specializations. Even today, research on smell continues to uphold some form of disciplinary disconnection.

Shepherd had found locally distributed activation patterns in the bulb in the 1970s, inspired by the English physiologist Sir

Edgar Adrian's work at Cambridge during the 1940s and 1950s.[31] Adrian, 1932 Nobel laureate in physiology, studied the firing of bulbar neurons with fine wire recordings. The responses were selective to different odorants: "So far then it looks as though acetone molecules will produce an excitation coming mainly from the front part of the organ and from the particular groups of receptors in that area which have this specific sensitivity to it." Adrian recognized that this localized pattern varied with additional factors, including the concentration of the stimulus, yet kept its core activation identity. "The result is that the electrophysiologist, looking at a series of these records, could identify the particular smell that caused each one." Adrian hastened to caution: "We must not conclude that the brain identifies the smell by the same criteria."

Shepherd knew of Adrian: "As I was finishing my doctoral studies at Oxford in 1962 I paid a visit to Adrian to discuss how my results might relate to his ideas about the representation of odors by spatial patterns of mitral cell activity in the olfactory bulb. What was the mechanism underlying the spatial patterns? I don't remember the details of our conversation, but I do remember his final advice: Look to the glomeruli."[32]

Shepherd picked up where Cajal and Adrian had left off. Into his hands played an optical imaging method newly devised by the biochemist Louis Sokoloff. This method, with its tongue twister name 2-deoxyglucose (2DG), targeted cells by autoradiographic imaging, which allowed for measuring regional metabolic activity in the awake-behaving brain. In 1975, Shepherd and his postdoc, John Kauer, working with a pioneer of the method, Frank Sharp, reported the first clear images of distinct local patterns of stimulus activation in the bulb.[33] It transpired that glomerular patterns also exhibited foci in activity, suggesting a core pattern in the processing of different odorant concentrations. Soon after, several other laboratories started taking up the trail of the olfactory map, and glomerular modules acquired the status of being a functional equivalent to receptive fields in the retina (Chapter 7). How should we interpret these bulbar patterns? More specifically, how does the olfactory bulb turn random activity in the epithelium into an orderly or systematic map? The neurogeneticist Peter Mombaerts, working with Axel, arrived at the answer to this puzzle in 1996.[34] Mombaerts et al. identified a remarkable genetic trick of the olfactory system. Recall that the receptor discovery revealed that all olfactory sensory neurons (mostly) express only one receptor gene, ergo a single receptor type. Experimentally, sensory neurons were thus acting as potential substitutes for studies about the functional behavior of the receptors. If one identifies the ligands that excite a sensory neuron, one also knows what ligands activate its receptor. Using genetic markers, one can trace to what place in the bulb the receptors communicate through these nerves. This is what Mombaerts did. He found that neurons expressing the same receptor gene all converged into one glomerulus, where receptor signals appeared as spatially discrete blobs. Such convergence of sensory neurons demonstrated that when a molecule activated a specific range of receptors, the signal resulted in a discrete pattern of bulbar blobs. A musk molecule causes a different activation pattern than a citrusy or fruity odorant. Each odorant yielded a unique pattern, like a fingerprint.[35]

Exploration of receptive fields in olfaction shifted from the epithelium to the bulb. Fifteen years after the receptor discovery, olfactory neuroscience held targeted access to the connectivity between the receptors and the brain.

Quo Vadis, Olfaction

In the mid-2000s, it looked as if the olfactory system operated by a topographic organization analogous to that of other sensory systems. And the spatial distribution of activation patterns in the bulb was expected to be maintained in the olfactory cortex. Nothing suggested otherwise for another decade. The pieces promised to rush into place at last. Less intuitive was what information was encoded in an odor map.

Topography in the bulb did not yield an order of odor. What was actually matched and mapped here? What features of odors linked to these neural correlates? The content of neural maps seems fundamentally contingent on a meaningful model of the input, echoing features in the environment. It might be intuitive to imagine how the brain uses neural space to encode spatial features of the environment in vision. This idea is not convertible to smell, however, as we explore in the remainder of this book.

Nearly everyone, from computational and wet-lab neuroscientists to molecular biologists, from neurogeneticists and sensory psychologists to flavor chemists, agrees that olfaction is far from being solved today.

Neuroscientist Mainland told how he entered the field: "You don't know the basics. You don't know, really, how receptors work. You don't know how intensity is encoded. All these things are known in vision. You have primary colors and you can predict what color you get when you mix them together—you can't do any of that in olfaction. There were a lot more open questions that were big and important, and so it seemed like a better place to study."

Jonas Olofsson, cognitive scientist at Stockholm University, emphasized the need for better theory next to experimentation in olfaction: "The philosophical and psychological understanding of what is the role of olfaction, I think, that's going to be critical to interpret the biology, also as an evolved, biological response. It's a function of the local environment. Understanding the properties of the context, the functional properties of the situation, without that we wouldn't be able to understand what kind of biological activities evolved and why."

"Now, are we data-limited or theory-limited?" the neuroinformatics researcher Rick Gerkin at Arizona State University questioned. "From a machine learning perspective, we're definitely data-limited because the machine learning model squeezes everything out of the data that it can. The only way it could be better is to have more data. From a theoretical perspective, though, it's still theory-limited. [The current model] doesn't tell you how olfaction works. So someone else could come around with an elegant theory that involves receptors and neurons and whatever and say: This is how olfaction works—from the same data."

The biophysicist Andreas Mershin at MIT shook his head: "This whole idea that we need a bunch of data to solve a lot of problems . . . that: 'Oh my God, if only I had enough deorphanized olfactory receptors, if only I had enough about the genes and primary segments of the structures, I would get somewhere!'— That idea is bankrupt. It has never been the case that, in the stage of science where we are right now, we're ever limited by data. There are times when we are limited by data, but in olfaction and in drug discovery and everything I do, it is not limited by the availability of more data."

Olofsson noted: "We certainly need better theorizing about the role of olfaction and how to interpret the axonal patterns—and why is it organized like that? I think that's something that needs probably a lot of theoretical questions, as well as empirical."

Olfaction is ripe for interrogation. What we need is to think about the data we *do* have. Mershin agreed. "In olfaction, you can talk about things without getting this big oppression of: 'Oh, no. We know exactly how everything works!' We don't know how vision works. It's a myth that we do. If I say that out loud—you can publish it, that's fine—but if I say that out loud here, people think that they know because there's lots of textbooks . . . olfaction is not as well represented, that allows a little bit of freedom to play around. But the system is completely generic in general for me. It's a model for neurobiology; it's a model for neurophysics. It's a way to understand emergent properties. It's a way to understand evolutionary biology. It's a way to understand this misconception about structure-function relations. And it's about phenomenology."

Its inherently cross-disciplinary outlook presents both a challenge and an opportunity for the field, Leslie Vosshall found, emphasizing the need to think beyond the schema: "And I am a reductionist!" She laughed. "I'm exactly the right person to be pushing back against [simplification] because I am, by training, a reductionist, trying to get these things down to the simplest number of pieces. But then all this stuff that's probably important gets thrown out . . ."

What may have been overlooked? For Terry Acree it is the subject of perception. "You cannot resolve the question of olfaction until you resolve the language about perception . . . You have to be able to say: What does it mean to be an odor? Is it a different level of the same odor? We don't clearly discriminate that in our language."

Current trends in olfactory neuroscience are opening up to a quintessentially philosophical investigation: What are we trying to measure and map onto neural space in the first place? What actually *are* odors?

CHAPTER 3

Minding the Nose

ODORS IN COGNITION

Ask six people working in olfaction what odors are, and you get six—if not more—answers. A chemist will tell you about the most minute details responsible for the odorous quality of a molecule.[1] See that hydroxyl group over there? The biologist is going to shake her head to emphasize that odors are more than chemical structures; rather, they are signaling functions in organisms.[2] Olfaction plays a behavioral role. The neuroscientist will nod before remarking that these behavioral functions come down to the activity of neurons firing in the brain.[3] At this point, the cognitive psychologist might intervene to say that such strict behaviorist or neural perspective is not enough to understand the mental mechanisms that embed odors in the cultural landscape of cognition.[4] Consider the role of memory, language, and learning in odor experience! A philosopher could add that while smells seem to convey mental images of things in the world, like apples or roses, their experience often borders on the fine line that separates conscious from unconsciousness perception.[5] Smells pass momentarily through conscious awareness as a fleeting and transient quality. How can we be sure that what we perceive is real? The perfumer will lean back in bewilderment, wondering why no one is talking about the distinct aesthetic experience and hedonic appeal of smell.[6] After all, you can tell a tale, entice a crowd with a fragrance. All of these viewpoints are correct; none captures the nature of odor entirely. Such a mosaic of perspectives makes it tricky to frame the discussion about the material basis of smell and how it connects to the characteristics of its perception. The nature of smell is not limited to a single scientific outlook. Maybe this is because the perceptual content of odor experience is somewhat twofold. Smell sensations are continuously directed at both inwards and outwards processes.

Certainly, you perceive *the odor of something.* Things smell. And most things smell unique. Remember your aunt's kitchen or your father's garage. Consider the variety of food aromas and spices across the globe, the sheer diversity in fragrant flowers and plants, even the characteristic odor of people and their body parts. Possible objects of smell seem infinite—not to mention the countless, sometimes outlandish creations in perfumery. It is clear that olfaction is a robust material experience; it tells us something about the world—something about people, places, and materials. The sense of smell provides us with vital signals about the invisible core of things.

Smells are not exclusively about things in the world. The experience of an odor usually does something to you as well. Smell sensation seems bound to individual evaluation. We engage with it according to several decision vectors: Do I like it? Is it pleasant? Is it (too) strong or intense? What is that odor; is its quality closer to a blackberry or a cherry? Would I eat it? Alternatively, is a particular fragrance more flowery and aesthetically pleasing? Would I put it on my skin? These numerous decisions involved in odor perception affect our experience of its quality.

The perception of a smell ties to an evaluation of its behavioral value, reflecting on your constitution as its perceiver. Perceptual judgment thus ties to physiological and mental factors. Think about it. Things tend to smell differently if you are hungry or sated, bored or engaged, joyous or enraged, pregnant or hungover. Sometimes you discover how hungry you are only after a sudden whiff of food aroma. Your nose conveys information not only about the world but also about yourself.

Olfaction is both exteroceptive and interoceptive—referring to a sense directed at phenomena outside and inside the observer. Philosophers refer to the exteroceptive role of olfaction as "intentionality," meaning smell is about things in the world. Food aromas tell you about the edibility and quality of a dish. At the same time, olfaction constitutes an inward reflection regarding the relative value assigned to this information. Food aroma makes you aware of whether you are, for example, desperate enough to eat ramen noodles for the third time in a day. Andreas Keller, a neurogeneticist and philosopher in New York, nodded: "We can respond very differently to the same smell depending on context, or to the same touch depending on context, whereas in vision that is not so much the case." (Although some vision scientists might disagree with Keller; do we also not just see whatever we want to see in some instances?)

Like no other sense, olfactory experience mediates observations of material changes in the environment in correspondence with our own physiological and psychological states. This is quite remarkable. Your nose grants access to extremely subtle chemical changes in the environment and, at the same time, relays the state of your embodied self, including the dynamic interactions between the two forms of experience.

Olfaction raises deep philosophical questions as part of the human mind. The lack of intuitive answers to these questions offers a welcome challenge to received philosophical views. It is an opportunity to revisit the theorizing that shaped our general understanding of the senses. What does it mean to "see" the world through the nose? What access does the nose grant us to material affairs? And what kinds of activity govern the formation of odor percepts? There are two ways of approaching this issue: we can start from cognition or behavior. In reality, both are inseparable. For clarity, let's begin by exploring the olfactory mind, and engage with the role of behavior in Chapter 4. There is much more to the mental life of odor than meets the "inner eye."

Two Senses of Smell

How many senses of smell do you have? "What an odd question," you may think. "One, of course." You have one nose, and you experience smells as one kind of modality. But a biologist—or kitchen chef—will tell you that you have at least two senses of smell.

Cooking has been central to human culture. The hunt for new flavors was the driving force of globalization: The spice trade shaped the socioeconomic landscape of the modern world like no other human endeavor.[7] Over the course of the twentieth century, eating was transformed by the increasing industrialization of food production and the discovery of artificial flavors.[8] Palate pleasers entice humans more than a Picasso.

Introspection, our "inner eye," conveys an entirely wrong idea about eating. Most people think food flavor is in the mouth, that it is an experience of tasting. That is not the case, or at least not exclusively the case. You eat with your nose. This idea sounds erroneous; don't you *feel* that taste is in the mouth? But "tasting" is a matter not solely of the taste buds. Think of the experience when you have severe congestion that blocks your nose. Everything tastes bland and uninteresting. Moreover, there are a proportionately small number of taste receptors in comparison with all the flavors you experience. Taste receptors on the tongue encompass food sensations of sour, sweet, bitter, umami, salty . . . and maybe fatty, according to recent research.[9] That's quite a limited palate given the abundance of flavor qualities. Where is the strawberry receptor on your tongue? Where are the minty, cherry, chocolaty, smoky, and garlicky receptors? There are none. These are all qualities created in the brain via the nose.[10] *The International Standards Organization* thus declared that "flavor" includes olfactory next to gustatory and trigeminal stimulation, and that "aroma" refers to retronasal smelling.[11]

Linda Bartoshuk, leading researcher on taste at the University of Florida, explained that your nose influences even intensity perception in food: "You can knock out one tympani nerve with anesthesia. When testing ordinary things like chocolate and ketchup, things you can buy at a supermarket, and you knock out the *chorda tympani* [branch of the facial nerve originating in the taste buds], the flavor of these common food items would drop by 50 percent! It's now half as intense flavor as before. If that's the case, then the interaction in the brain between retronasal olfaction and taste is one of the most important phenomena that exist."

How do you taste with your nose? There are two pathways for the detection of odorous chemicals. The process of *orthonasal olfaction* takes place when you sniff and inhale—which is what we usually think of when we talk about the sense of smell. Then there is the process of *retronasal olfaction* that happens when volatile molecules travel through the back of your throat up to your epithelium. Look at Figure 3.1; here you see that an opening, the pharynx, connects your mouth and nose. This open space allows the food molecules, released during chewing in your mouth, to get to the back of your throat. With the upcoming warm air from your lungs, these inner-mouth flavor molecules reach and interact with the nasal epithelium after swallowing. Swallowing acts like an air pump connecting the lungs to the epithelium. Test the mechanism on yourself! Hold a finger right in front of your nostrils. Now swallow. What you feel on your skin is a gentle flow coming out of your nose. That is the air from your lungs.

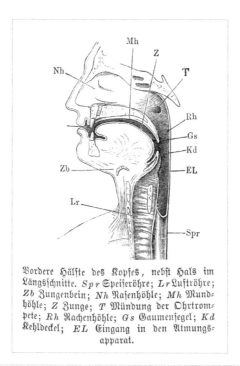

Vordere Hälfte des Kopfes, nebst Hals im
Längsschnitte. *Sp r* Speiseröhre; *Lr* Luftröhre;
Zb Zungenbein; *Nh* Nasenhöhle; *Mh* Mund=
höhle; *Z* Zunge; *T* Mündung der Ohrtrom=
pete; *Rh* Rachenhöhle; *Gs* Gaumensegel; *Kd*
Kehldeckel; *EL* Eingang in den Atmungs=
apparat.

Figure 3-1 Historical drawing showing the anatomy of retronasal olfaction. The
pharynx (Rh) connects the mouth (Mh) with the nasal cavity (Nh). Warm air
from the trachea (Lr) reaches the pharynx and catapults food odorants up to
the nasal epithelium where they interact with olfactory receptors situated on
the sensory neurons. Source: Quagga Media/Alamy Stock Photo.

This retronasal pathway distinguishes humans from most animals, even those with a good sense of smell. Mice do not have a retronasal pathway; neither do dogs. These animals are fantastic at
sniffing. But they would never know what it is like to taste great
food. Their world is somewhat comparable to your experience of
coffee, with its great orthonasal smell, luring you out of the cave of
your bed in the morning. Wonderful smell notwithstanding, the experience ends up with a disappointing taste: coffee mainly tastes
bitter. (Retronasal aroma does not necessarily match the qualities

of orthonasal scent. We will get to that in a minute.) What enables humans to experience flavors when so many other animals cannot? The difference between humans and many other animals, allowing for retronasal smell, is a missing bone.

During evolution, humans and other primates lost the bony transverse lamina in their noses, resulting in the morphological differentiation between what is called the primary and the secondary nose.[12] This bone divides the olfactory and the respiratory tract in "macrosmatic" (having a highly developed sense of smell) animals such as dogs. One result of this development is that olfactory recesses in humans are no longer separated from respiratory ones, creating the nasopharyngeal space. The disappearance of the bony transverse lamina turned the mouth into a second nose. Whereas dogs and mice have one nose to smell stuff, one could say we have developed two!

The second nose, affording retronasal smelling, is what makes food taste interesting. It also is what makes cheese bearable, even delicious. Have you ever wondered why people eat stinky French cheese despite its, well, awful stink? Or why coffee smells excellent but tastes disappointing? What you experience *orthonasally* can differ in its perceptual quality from what you receive *retronasally*—considerably so. This difference is partly physiological; it is due to the varying warmth of air from the lungs as well as airflow. Airflow determines which molecules reach your epithelium first, and at what speed, so much so that what the receptors recognize in each case differs in activation patterns of temporal coding (Chapters 5 and 8). Accordingly, the same odor source can elicit considerably different qualitative experiences depending on the physiological route of access—a fascinating phenomenon, impossible to explain without the sensory system making sense of the stimulus. The experience of cheese via the orthonasal or retronasal route differs. Both perceptions are accurate sensory expressions of their target: the cheese.

The hedonic note of an orthonasally or retronasally delivered smell also varies. Stinky cheese is more pleasant when you eat, rather than sniff it; conversely, the coffee aroma is better than its retronasal flavor. This divergence might connect to implicit reward experience. Coffee could taste so disappointing because the sophistication of its flavor does not match its highly suggestive orthonasal aroma. Cheeses frequently taste better with further experience of cheeses, and their retronasal character holds pleasant notes that are not revealed by their orthonasal aroma. Recent studies by the psychologist Dana Small and others have shown that neural responses indeed differ for the orthonasal and retronasal processing of odorants.[13]

Why do we feel that flavor is in the mouth, not the nose? The knowledge that flavor originates in the nose does not change its phenomenological placement in the mouth, a phenomenon called oral referral.[14] You cannot cognitively penetrate and redirect the oral referral of flavor, however much you try. That's because the mouth is where the action happens. Your awareness, as integrated sensation, is actively bound to the locus of voluntary action. Terry Acree noted: "It is *what happens* in your mouth. You really see that when you study wines—the mouthfeel, the texture, the interaction between the components, and the sweetness, and sourness . . . All those processes in your mouth have to be right for you to be happy."

Besides, if you detect toxic or rotten elements in food, quick reactions are essential. "It makes biological sense because if that flavor is something you need to spit out," Jay Gottfried, neuroscientist at the University of Pennsylvania, laughed. "If you think that flavor is in the nose, you may be tempted to snort it out."

Retronasal olfaction involves physiological determinants different from those of the orthonasal pathway, like warm air from the lungs. Physiology contributes more than the nuts and bolts that underpin sensory perception—it explains what smell perception actually is. Attention to physiological detail adds more than empirical data backing up our understanding of perception. It challenges our underlying theoretical assumptions. The discovery that phenomenological experience might not match (or explicitly mismatches!) its real material causes calls for new philosophical thinking, further touching on conscious access to the senses. Bartoshuk pointed out that observability and introspective access in perception links to our experience of activity. Tasting is voluntary action in taste perception while smelling is the activity verb connected to inhalation in orthonasal smelling. There is no activity verb for flavor as a retronasal experience since there is no voluntary action involved. Because flavor is *experienced to happen* at the same time as the conscious activity of tasting, and not *experienced to be undertaken* as an independent movement, it gets experientially associated with tasting. And so introspection can mislead us even about the type of experience we have.

Nonetheless, the experience of flavor itself is intentional; it does tell us something important about things in the world. The philosopher Barry C. Smith at the University of London emphasized: "One distinction should be clear. If there is such a thing as flavor perception, then perception is also perception of something. We want to say that our perceptual state is either accurate or inaccurate, that you're correctly perceiving or misperceiving. But if there was nothing we could say was flavor, independently of having something that it was supposed to be *portraying as flavor*, it wouldn't be answerable to anything but itself. That would be a state which didn't have any representational powers. It was just a state which existed whenever flavor experiences were created by the brain. This would be rather like having an itch or having a pain."

From this perspective, philosophers would say that the sense of smell (retronasal as well as orthonasal) has representational power. For perception to be representational means that it expresses a mental image that, in some way or another, relates to external features in the world, implying that there must be some form of success or accuracy condition. Success conditions are not easy to define, however, especially in olfaction. What makes a particular odor perception an accurate representation of smelly stuff? After all, odor experience routinely varies from observer to observer. How could odors act as representational expressions of reality if perceivers lack a sufficient degree of consistency in their responses to smells? Answering this question requires us to consider the roots of variability.

Smell in Conscious Perception

Olfaction and its link to cognition were long disregarded due to philosophical bias that treats perceptions as conscious phenomena with propositional content. Such bias excludes many fundamental perceptual capacities, nonconceptual and nonconscious, capacities of high relevance for the analysis of odor object formation.

Conscious perception is routinely analyzed in the form of "conceptual content." The propositional beliefs we hold about the world express the conceptual content generated by the senses. Such content entails variations in sophistication; for example, we say, "These two socks have a different color," or "The beginning of Wagner's *Flying Dutchman* signifies an early instance of the breakdown of tonality." While the level of sophistication of our beliefs may differ, the propositional attitude of their content is similar. In its propositional form, perceptual content underlies correctness conditions that link to features of physical reality. "These two socks have a different color" is correct if these two socks indeed have different colors, as measured by the sensory representation of color via computation of its electromagnetic spectrum.

Talking about odors in this way is not straightforward for three reasons. First, it is not intuitive at all what would make an odor perception an "accurate representation" of stuff in the world. Say you know what coffee smells like. In coffee aroma, you do not perceive the presence of indole, a molecule with a strong fecal quality.

Indole is a component in the overall plume of coffee aroma. Is the absence of indole in your perception an inaccurate experience? Before arguing that the lack of indole in coffee perception is the proper mental representation of its aroma, ask any pregnant woman with heightened olfactory sensitivities to certain odors. She likely perceives it. Whose perception is "correct"? The question appears as ill posed.

Representational accuracy is not the only puzzle. It is not even evident what we are consciously aware of when we smell, let alone what the conceptual content of odor is. Recall the instances when you smell something without seeing its source. Often you are unable to tell what it is. Judging what an odor is can differ profoundly if its source is visible or invisible. Even when we cannot seize the conceptual identity of an odor sensation, we still smell something distinct. This prompts the following question: What is the mental representation of odors really about?

Last, you may also encounter sensations that, although they are caused by odorants, you do not recognize as being odors— remember food aromas.

People underestimate the impact of smell on their mental life and decision-making. Most people, when asked which sense they'd give up, opt for smell without hesitation. This sounds reasonable since modern civilization relies heavily on information processed by vision and audition. But the general disregard for the nose is worryingly high and often misguided. A 2011 survey by the McCann Worldgroup asked young people whether they preferred to give up a technological gadget, like their computer or cell phone, or lose their sense of smell.[15] More than half of the participants between sixteen and twenty-two years of age voted against their noses, a number perhaps higher today. Hold that thought just for a minute. A significant number of young people would disable themselves permanently rather than give up something as replaceable as a technological device.

Why do we underestimate the sense of smell? We seldom are consciously aware of it. And here's the caveat: Smells are not always in explicit conscious awareness—but such unawareness does not mean that olfaction is not integral to conscious experience. Our mind might not knowingly track and attend to odors all the time. That does not prevent olfactory influences from modulating overall conscious experience, including other modalities. Smith heartily agreed: "Our odor processing is continually giving us a modulated experience—kind of *background to consciousness,* against which we register change and events, and against which we pick out things that have an impact on our emotion, our memory, our search for food, our attraction or repulsion to people, places, and things."

Olfaction is not always part of what philosophers discuss as "conscious awareness," an explicit state of mind that can be referred to ostensively (like "this red color of the ball" or "this high-pitched sound"). Smell is frequently embedded in the conscious experience of the world without being foregrounded as an olfactory experience. When you sense the atmosphere of a place as being sticky, fresh, or feeling narcotic, you might not linger on this impression as olfactory. It is just one part of your general, multisensory perception of a scene that also includes indistinct sounds and cloudy skies. Gottfried added: "I heard a philosopher speak about how losing the sense of smell was a kind of constriction of time and space. For example, if we could smell the seawater right now, that gives us a greater sense of the world. She felt very trapped within time and space than if she had access to her sense of smell. That was a really eloquent way of framing things."

Try recalling the last five scents you consciously encountered today. Can you honestly say that you remember the last five, even three smells of which you were conscious, much less in the right sequence? Did you first smell the coffee or toast for breakfast? You possibly now run through your eidetic memory of this morning,

not your olfactory one. You rarely remember your olfactory encounters, unless they were prominent. A perfumer or another
smell expert may remember because explicit olfactory awareness
is part of their daily cognitive landscape. For the ordinary person,
who pays little attention to olfactory encounters, recalling even the
last odor she was aware of poses a challenge. It is not a principally
biological constraint we are talking about here, but how we atten-
tively engage with the world as conveyed by the senses.

It is a matter of attention in memory formation and the conditions of comparison. This fact does not separate olfaction from vision. Keller argued: "It's a memory problem. That's been proven in
vision over and over. If you have a visual display and then you
change it, everybody goes: 'Oh, it changed.' But if you have a visual
display and then you have a mask, people have a hard time finding
that change because they can't remember that entire scenery
they've seen and so the difference is more complicated." Still, unlike smell, we always are consciously aware of something in our visual field. The same cannot be said for human olfaction. Odors,
in particular, seem to border on this fine line between conscious
and unconscious perceptual processing.

How does unconscious perception shape conscious experience?
Many olfactory stimuli in the environment never reach a threshold
or detection level. Still, stimuli at subthreshold levels (which we do
not perceive consciously) influence the conscious perception of
compounds above the threshold level. Subthreshold influences on
perception are measurable, although challenging to implement,
and are a genuinely fascinating phenomenon. Subthreshold forces
are not limited to intramodal effects (between olfactory stimuli
of different character or intensity); they also affect cross-modal interactions. Imagine your sensory systems like switches on a control surface, like in a recording studio. Amplifications of one modality will affect the overall perceptual complex, so that other
modalities need augmentation, either by turning them down or

through intensification.[16] Your sense of smell is a vital switch in your mental recording studio.

Acree highlighted: "We have lots of experimental evidence that subthreshold molecules can affect suprathreshold molecules [molecules in a concentration range large enough to pass perception threshold], how they perform. There are numerous experiments that are credible that say that subthreshold levels of all kinds of stimulants will change the response to suprathreshold limits of other stimulants." These findings fit with studies on the visual and auditory system that likewise found subthreshold influences on conscious perception.[17]

"Which is why the whole question of what is conscious becomes central to the issue. Because if we measure behavior, we can interrogate subjects about their conscious experience." Acree continued. "We can't, in some sense, interrogate them about their nonconscious experience. Which has to be enormous if you consider that in a mixture a lot of the chemicals that are above their threshold are all firing all these receptors. A lot is going on that doesn't rise to the level of consciousness. Or it may affect what does rise to the level of consciousness."

That poses a challenge to in-depth studies of odor experience. Notwithstanding, experimental research on olfaction has reached a point at which it (finally!) can link variations in smell experience to physical causes—for example, to receptor genetics. Which stimulus reaches threshold level, and at what concentration, is contingent upon receptor expression and sensitivity. People have individually variable patterns of receptor expression in their noses, and so your pattern of receptor expression differs from mine. Besides, there are diverging sensitivities to smell. It thus stands to reason that an odorant—interacting in a combinatorial fashion with a range of receptors—could be perceived differently in people with a distinct receptor repertoire. That is in principle testable.

Two recent studies (involving Andreas Keller, Leslie Vosshall, Hiroaki Matsunami, and Joel Mainland) found that variation in the genetic basis of olfaction resonates with interpersonal differences in odor perception.[18] Genetics is closely linked to smell. Additionally, the vast number of olfactory receptor genes in the human genome allows for many mutations to take effect. Sometimes, such mutations lead to notable differences in the experience of odor. As an example, some people really dislike cilantro (coriander); they perceive its aldehydes as soapy and pungent instead of fruity and green. This difference in perception is due to a genetic variation next to the olfactory receptor gene OR6A2.[19]

Such examples show that traditionally philosophical interests, such as the nature of qualia (wondering whether my mental representation of the color blue is the same as your mental representation of blue), now link to empirical examination in olfaction. We both may call it "rose," but your qualitative experience of a given rose blend will differ from mine if our receptor repertoire varies. (How qualia diverge, or what this means on a mental level, is another matter and much harder to evaluate. Partly, this is because nobody seems to agree on what "qualia" actually are, or, for that matter, whether they exist.)[20]

Biology is not the only discipline that shapes perception. There is also psychology. Contextual and learned behavioral influences are essential in this context. Cross-cultural studies on sensory performance, for example, have documented the fact that familiarity with an odor influences its threshold level: the more familiar you are with a smell (say petrichor), the lower your threshold for its detection.[21] Olfaction exhibits a lot of variation among individuals; these variations are not arbitrary. They link to fundamental causal factors and mechanisms, and these factors can be studied today.

Olfaction forces us to move beyond "the myth of the given." Smells are not presented to the mind for immediate contemplation or rational integration into the cognitive architecture. On the contrary, it sometimes requires considerable effort and conscious reflection to reveal the hidden, yet extensive, presence of smell as part of our inner mental life. Some odors are recognized only after paying attention to our sense of smell, and it takes considerable effort to analyze the content of your olfactory experience through directed attention (Chapter 9).

Gordon Shepherd remembered an exchange with the sensory scientist Avery Gilbert, a colleague of his: "I began to realize that there needed to be something done to make people realize that we're not so bad at smelling. I wrote this little article on the human sense of smell: is it better than we think?"[22] Shepherd smiled. "Avery then sent me an email and said: No, you've got the title wrong. It should be . . . 'The Human Sense of Smell: Is it better *because* we think?'"

That leaves us to wonder, what are we aware of when we smell? What does the mind-brain do with information from the nose? And what is such information for? This first foray into the invisible, sublime life of odor is relevant to keep in mind, literally, since what is visible to the conscious mind reveals only fragments of its mysteries. Not all olfactory sensations entering conscious experience are treated as conceptual objects, or even explicitly as olfactory objects. That does not make them less substantial to conscious awareness. As sensory background to conscious experience, olfactory perception actively contributes to the overall cognitive representation of scenery. But at the same time, the conscious awareness of smells requires rethinking.

Odors as Objects of Cognition

The conscious experience of smell defies coherent, uniform description. "Odors are so fuzzy to us in perception. In vision, you can see an object, follow an object, and perceive the object. Odors, they sneak in. All of a sudden, they are there. Or you can follow them. But typically what happens with an odor is: 'Oh, something smells!' But it's already there. I've been noticing this for quite some time, but it's only now that I get conscious about it," said Johannes Frasnelli, a clinical scientist at the Université du Québec à Trois-Rivières. In their spontaneous conscious appearance, scents can blindside you. Are the objects of olfaction harder to conceptualize because the sense of smell (unlike vision) is not continuously at the forefront of conscious mental life?

It is easier to judge whether two scents are the same than it is to name them. Sometimes you find yourself wondering, what *is* that smell? The semantics of odor imagery is constitutionally ambiguous. This is recognizable once the source of a smell is invisible or unknown. Perceptual content in olfaction is influenced but not tied to visible sources.

An odor given on its own can evoke substantially different imagery and associations than when it is presented with a visible source. Hans Henning discovered this inherent contextuality in odor semantics as early as 1916 (Chapter 1). Henning tested the impact of visual cues on odor identification. This led him to introduce a fundamental methodological distinction: between "the true odor (*Gegebenheitsgeruch*), which is obtained by the observer who is smelling with closed eyes and is ignorant of the nature of the scent, and the object-smell (*Gegenstandsgeruch*), which (like color) is projected upon the objects from which it is known to come and apt to be distorted by associative supplementing."

The neuroscientist Noam Sobel, at the Weizmann Institute of Science, likes to demonstrate this distinction to skeptics with a simple test.[23] You need only go to the kitchen. Next time you go to your fridge, smell its ingredients with a blindfold. Ask a friend to pick ingredients for you to sniff randomly, and try to name them. It is surprisingly hard. These are things you know, stuff that you bought and put in the fridge yourself. So it is not merely a matter of being familiar with a specific odor and its quality. This experience yields dissociation between two kinds of processing: odor perception and odor naming. You do smell *something,* and you can tell that something apart from the smell of something else. It is just so darn tricky putting a name on it!

So the causal objects (odorants) are not mere mediators between the source of an odor and the semantics of its mental image. This is of practical consequence, too. The clinical scientist Thomas Hummel at the Technical University Dresden remarked: "We have to be aware that this is an issue when we clinically test with smell identification kits." These tests "normally present you with a smell. You sniff it, and you really have trouble identifying it. Even if it's a really popular odor like pineapple. This is why the trick is not just to present the odor, but also to offer a list of descriptors. In the list of descriptors, there would be, for instance, tar, grass, leather, and pineapple. And then it's easy for you to identify pineapple if you have a sense of smell."

The troubled relationship between scent and language is often remarked upon and has fostered philosophical and public disinterest in the cognitive basis of olfaction. Its complicated ties with naming may constitute the chief reason behind the outright denial that olfaction has a cognitive foundation, making it intelligible only as a brutish sense of instinctive sensations. This is because language, if anything, acts as the observable—visible and audible—expression of the human mind. Its semantics are the medium and mirror of cognition, volition, and intention. Something that language cannot capture stands unyielding and is tricky to study. The popular belief about our inability to describe odor needs an overhaul.

The common tenet is that we lack an adequate vocabulary for olfactory sensations. That is not true. Consider a comparison with color. Of course, we can name the basic colors of the rainbow. But there are others you may have never heard about unless you are a color expert: Mikado (a bold yellow), glaucous (a powdery blue), Xanadu (a grayish green), and many more. That parallels the case with smells: we have common names for general categories (flowery, fruity, woody) and more specific terms that require training to use (blueberry or ylang ylang). It is not merely a matter of having a language but making good use of it. Besides, consider all the flavors (retronasal smells) you can name!

We struggle to define the distinctive quality of smells without specific training. But some complex visual objects, such as faces, are equally hard to describe. Shepherd replied: "The problem of putting words on odor perceptions and flavor perceptions is difficult because that's exactly the problem that we have with any visual pattern we see. It's irregular, and that's why I use the comparison with faces.[24] We have no vocabulary for precisely characterizing someone's face to a stranger who doesn't know that person because these are irregularities that aren't . . . we can't say that's vertical or horizontal or whatever. It's just, well, there're two eyes, there's a nose, and a mouth. But how do you characterize that in a way that will enable you to identify somebody amongst a thousand other faces? It's almost impossible. And that's the same way with smells." Vocabulary as such is not the issue.

What discerns the cognition of smells from that of visual objects is the challenge of linking olfactory sensations to proper names—that is, unique identifiers. Even if we come up with descriptions of olfactory sensations, they differ considerably between people. The psychologist Theresa White at Le Moyne College connected the dots: "Truthfully, people are miserable at producing unique identifiers. They cannot put a name on it, but they'll call two or three things in a row licorice, and they're really diverse. Perhaps, one of the reasons for this is that in order to perceive an odor, you need many receptors, some of which are much more active than others. Some of the receptors are used to perceive multiple odorants. Maybe this lends to some of the confusion as to exactly what you're dealing with, so that you can't tie a tight label to it. If you think about all of the oranges that are out there that we have one label for, but things that are sort of tangerine, you might still smell as orangey. Things that are almost lemony, you might still smell as orangey. Whereas for many types of stimuli [in other sensory systems], we have a pretty tight distribution as to what we would give to that particular label. For odors, it's a pancake shape. It's pretty broad, and I think that mimics the number of receptors involved. I think that is some of the difficulty with the language." Such underdetermination in odor coding, from the receptors onward, indeed might link to ambiguity in perceiving odors (Chapters 6–9).

People are known to be notoriously bad at giving odors unique, generalizable lexical identifiers. Is this a somewhat intrinsic, biological feature or more of a cultural factor of neglect in the cognitive processing of smell? At first, this phenomenon looks like an economic trade-off in evolutionary development. One plausible scientific explanation notes that olfaction and language processing share some cortical resources in neural coding. That implies that olfactory performance in humans could have diminished as a result of increased language processing. General difficulties in the attribution of words to smells count as additional evidence for this trade-off hypothesis. Impoverished linguistic abilities in naming odor would mirror a general condition of human biology. As White added, "Tyler Lorig would say that there are brain limitations in terms of tying the specific odor to language."[25] It is as yet uncertain whether this explanation is conclusive.

Anthropological studies suggest that the limited scope of odor language is not a biological fact. It might be grounded in cultural or behavioral indifference. Some societies work with a profound lexicon of smells that relates to a richer culture of olfactory habits and customs, as the cognitive scientist Asifa Majid (previously at

Radboud University, now at the University of York) emphasized. Her research on language also involves South Asian communities. "For the Jahai," a population indigenous to Thailand, Majid explained that smell is of utmost social importance.[26] "You can't cook two different types of meat on one fire. The Jahai will make two separate fires some distance apart so that the smell of the meats doesn't mix—because if they mix, their thunder god in the sky smells them and that's bad. You're breaking a taboo if you do that. Or certain smells make you sick, but you can use other smells to counteract them. Or a brother and sister shouldn't sit too close together because their smells mixing is a kind of incest and so that shouldn't happen."

The cross-cultural examination of language and behavior can provide fresh insights and attract broader attention for the conditions connecting perception and cognition. Majid's focus on cultural differences in language originated from her interest in the principles of cognition. "I'm interested in how the mind works. The way that I approach that is to look at a diverse cross-linguistic, cross-cultural sample to see what's similar across diverse cultures and what's different, and what might that tell us about what's uniquely human and what's shared across people. I started thinking about what language was bad at as well as what it's good at. Often we try to find those things that language is particularly good for. I thought it might be informative to think about where does it fail, where do we struggle, and what might that tell us about what language evolved for and how it hooks up to our perceptual systems."

In addition to the Jahai, the breakthrough was a 2014 study Majid published with her graduate student Ewelina Wnuk that documented the odor lexicon of the Maniq.[27] The odor terms of the Maniq revealed abstract categorizations of odor sensations, not just terms—like garlicky or rosy—derived from visible targets. These categorizations did not relate to source objects, but qualities, encompassing different kinds of objects under an umbrella term. Could the rules of cognition—how perceptual objects turn into conceptual objects—be culturally mediated?

Opinion about the interpretation of these results and their generalizability is divided. Jonas Olofsson noted, "There's a lot of ways to slice this issue. It would be a fundamental misunderstanding to say that learning a culturally mediated odor language is contradictory to the notion of universal language constraints." He added, "Even we have not said that our inability to verbalize the smell is somehow . . . that there's a biological cap on what we can do.[28] But it seems to come harder for us than to verbalize other kinds of materials, like visual material. This limitation, like any biological limitation, is subject to quantification throughout our lives." The ambiguity and variability of odor language may not be a cognitive limitation. It could be a consequence of how odor coding is distributed at the receptor level, as White suggested.

Besides, there are relevant differences in how people use odor language, Olofsson added. "We have smell experts in our own culture as well. They're amazing at describing smells. There's no universal limitation on odor language, but there might be a restriction. This kind of linguistic knowledge, it's harder to achieve. I think you have to see those as two separate issues in a way—learning and that kind of biological limitation."

Understanding of linguistic abilities in olfactory learning must take into account the fact that odor vocabularies are diverse. Even expert vocabularies are inescapably domain-specific, given the vast range of odor and flavor materials. Aroma vocabularies of wine tasters, for example, vary from those of beer or coffee experts. A significant difference is the category "bitter." (Strictly speaking, bitter is a taste; it helps to illustrate the point.) Bitter in wine tasting is a feature of bad wine. In beer and coffee, it is a qualitative note that distinguishes various styles and brands. Here, the categorization of bitter invites specification, so that coffee and beer experts work with different distinctions and categories for bitter. Wine tasters skip this step. They would not know one subtle bitter note from another because they are not trained to draw perceptual distinctions in this category. Domain-specific vocabularies also vary markedly between practitioners even within the same field. The expert winemaker Allison Tauziet at Colgin Cellars in Napa Valley confirmed: "We're using all these descriptors about the wine all the time. We've honed our own dialogue. But if I were to walk into another vineyard, then I'd be like . . . What the heck are you guys talking about?" Odor language is strongly contingent upon the rules of conventionalization. Majid concluded, "If you have conventionalized vocabulary, then it's something that speakers in the community all share."

It is far from evident how language use connects perception to cognition and vice versa. And so we saw how the intellectual attitude, which has traditionally sidelined olfaction as a cognitively unsophisticated sense because of the descriptive challenges in expressing its experience, has followed three mistaken assumptions: that introspection is a solid source for the analysis of perception (which was not the case in flavor), that conscious experience accounts for the structure of perception (which ignores unconscious influences shaping this experience), and that language provides a mirror of experience (ignoring that verbalization is a product of cognition and culture that does not provide a "separate tool" for the analysis of conscious perceptual content). All three assumptions invite a revision of our understanding of odors as cognitive objects.

Odor perception cannot be sufficiently accounted for by the "mind's inner eye" and its expression in propositional format. Focusing on semantics is deceptive. Semantic labels are not unambiguously representative of the cause of an odor. This point is more than philosophical; it also impacts experimental design. Vosshall argued: "I think that the field has to get away from words. We have to stop using semantic descriptors for things." Substituting words for perceptual content overlooks the contextuality of semantics, historical and cultural. Vosshall referenced Andrew Dravnieks's *Atlas of Odor Character Profiles*, a compendium of odor descriptors from 1985, which is still routinely used in olfactory psychophysics.[29] "Those 146 descriptors. . . . Fifty years from now, half of the words will be incomprehensible to people because they're very specific, culturally based in products and ingredients. It's like reading a cookbook from the 1500s! You have no idea what the ingredients are, or what the style of food is. You could never . . . without having someone come back, travel through time to tell you what that dish is or what it tastes like."

That is not to say that language use doesn't tell us something about the cognitive processing of odor; it is merely that the psychological roots of odor experience creating its content cannot be inferred from their verbal expression. "I think the future of psychophysics is not semantics, but looking at how people make judgments about how similar a molecule is and if they can discriminate it," Vosshall concluded.

Words are useful as cognitive handles in the communication of inner experience. But the process of odor categorization is not the same as odor naming. Gilbert shook his head: "As if the verbal characterization is somehow encoded in the orbital structure of the molecule. It's just wrong, philosophically." Gottfried replied: "Humans can't avoid naming things. There's a fascinating philosophical tradition around the object, and the name, and reality and all that stuff. Magritte had it so nice: This is not a pipe!"[30]

Is the alternative a more materialist perspective, centering definitions of perception on the target of odor experience? After all, smell perception is the response to something outside of us, an invisible world of molecular information. Matching the content of perception to its material origins comes with its own caveats, however.

The Intentional Target, Semantic ≠ Causal Objects

It's far from intuitive what the intentional target of odor perception is. Vision, as the paradigm sense, is primarily framed in terms of object recognition. Visual objects have shapes and colors, and we can check whether our mental representation of an object correlates with these features. In audition, we hear events associated with objects by temporal distance. Touch gives us a sense of immediate interaction with foreign bodies. These sensory encounters yield features generally associated with objecthood, such as material presence, location, and boundaries or discreteness.

Do odors comparably communicate material objects? If so, which ones? Three options for potential objects of odor perception are available, according to the philosopher William Lycan, University of North Carolina at Chapel Hill.[31] First, there are the things that give off volatiles, the macromaterial objects like the goulash boiling on the stove. Second, there are the volatiles themselves: odorants with their own causal characteristics (such as airflow turbulence). Third, there is a mental image, like goulash smell or rose odor. The connection between these three levels of analysis is not linear, but cross-cutting (Chapter 9).

Its material basis explains a lot of the peculiarities of odor perception. Molecules travel considerably more slowly than light photons do. After the macrosource is long gone, some of the volatiles remain: cold smoke lingering in the room and in textiles after a cigarette break, the pervasive clouds of heavy perfume, or the remnants of food aromas after dinner. The nose can signal sensations to your mind that are not bound to a visible source. Additionally, your sensation might stem from various objects. Think about stinky cheese, a piece of Époisses. (Its aroma is so strong that it has been banned from public transport in Paris.) Another source could cause the experience of this smell. There need not be any cheese at all. An evil perfumer may have created a synthetic mix to shock customers or hand in her resignation! Inference from molecules to perception or, conversely, from mental image to material object, is a misconceived conception. The causal objects (odorants) are not mere mediators between the source of an odor and the semantics of its mental image. This idea circumvents the processes that explain how the mind-brain creates odors from molecules and what odor images are. It further misrepresents the role of the stimulus as the cause of perception. The content of odor perception is not explained by reference to the microstructure of odorants.

A comparison with vision will help. Color vision is a calculation of wavelength. Smell, on this account, presents a sensory expression of the chemical composition of the stimulus. Similar to an optical physicist associating colors with the electromagnetic spectrum, it is plausible to think like an analytic chemist and assign groups of odor objects to their molecular basis, looking at functional groups, benzene rings, and double bonds in connection to odor quality. The philosopher Ben Young at the University of Nevada, Reno, entertained this idea as the "Molecular Structure Theory."[32]

Unless you are a chemist designing synthetic compounds, this won't help you to understand olfaction, let alone provide a model to analyze the content of odor perception. The relation between odor quality and stimulus chemistry is exceedingly complex. Explanation of odor via stimulus chemistry requires a model of the system that encodes the molecular information. This principle likewise applies to vision or any other system, for that matter.

From a systems theoretical perspective, stimulus coding in olfaction differs from the visual system in some critical respects.[33] In color vision, the wavelength is chipped into uniform ranges via the receptor cells, allowing for discrete computations of color categories and consistent assignments of perceptual qualities to physical features. Olfaction does not afford an equally linear structure-odor correspondence: odor types do not yield structural homologies between odorants. Considerably different odorants cause comparable perceptual qualities and, vice versa, similar odorants produce substantially different olfactory qualities—this point relates to the coding principles of the system, not the stimulus in isolation. The stimulus is the cause of perception, and the perceptual image is its content. How these two factors hook up is the thing in question.

The visual system teaches us that much as well. You won't understand color as wavelength without a model of how its sensory system works. Consider pink. The color pink has no associated electromagnetic spectrum, and so we cannot experience pink as the quality of a physical object. Pink perception is a concoction of your brain. Specifically, it is a sensory expression of a physical impossibility. It sounds paradoxical, but when we see pink, we perceive a stimulus "gap." This gap refers to your brain doing the sensory calculation of "white light minus green light." The color pink is a mental result of our mind "coloring in" for stimulus absence, such that the color pink is a neural computation of physical features in the world instead of a direct representation of them. In other words, it is a computational, not a physical, feature. The physical basis of perceptual content must be parsed from the viewpoint of the coding system.

A great example to illustrate the perceptual computation of odor is Sobel's discovery of "olfactory white."[34] Olfactory white is a nondescript odor quality created when you mix thirty-plus molecules with diverse, nonoverlapping chemical features. The remarkable thing about olfactory white is its quality. It has no associated ordinary semantic object (like apple). Its smell is not encountered in nature. It is meaningless to ask to what kind of object its content corresponds since there is no ordinary object available.

Moreover, olfactory white has no unique or definite microstructure. Note that you can blend any assembly of thirty molecules as long as their microfeatures are nonoverlapping. Olfactory white is not caused by a specific set of molecules or molecular features. The brain creates the perceptual quality of olfactory white when it is forced to cope with an overflow of physical stimulus information. The brain creates olfactory white when it does not know what else to do, just like white in color vision. In the case of vision, the brain faces the entire visible light spectrum, an overload of information it cannot cope with. So it creates white.

Olfactory white and visual white are computational features created by the coding system.

The experimental study of olfactory white is not easy. The mixture must contain odorants in equal intensity. But varying evaporation rates of volatiles soon alter their intensity and differentiate different combinations of olfactory white, changing their perceived quality. Christian Margot explained the details. Substances evaporate. "Any volatile substance evaporates under normal conditions. In a closed system, the equilibrium partition between liquid and vapor is constant for a given pressure and temperature. Conveniently, this constant is referred to as the vapor pressure. In the normal experience of fragrances and aromas, the equilibrium is rarely attained, because you always have an open system where some draught is stealing your molecules or diffusion enabling them to leak away." The evaporation rate is a critical determinant. "Evaporation rate may have some relationship with vapor pressure and enthalpy [a calculable property of a thermodynamic system] of vaporization. There are additional things that matter—because the rate is not about the equilibrium, but how fast you reach equilibrium."

Margot pointed at the beer in front of him to show how chemistry governs things in daily life. "The alcohol is held captive by the water. So it has a hard time, but it tries to continuously build the equilibrium between the liquid phase and the vapor phase. It collaborates with some water. But since it's not a closed system—there is wind, and sometimes I'm tipping the glass—the alcohol phase goes away. So the process of evaporation, the flux, is highly influenced by what is present in the liquid phase—also the interactions, and what is in the gas phase. If I had the alcohol in paraffin medium, there would be minimal interactions. The alcohol would go away rather easily. It's the same with fragrances you have in a mixture. You have components that may interact strongly, and others that will repel each other. So if you have oil, it will repel the alcohol and be happy. Just go away, leave me in my hydrophobic medium alone," he laughed.

Thomas Hettinger, a chemist at the University of Connecticut, added another challenge: matching intensity between mixtures. "Matching the intensity is a very difficult thing to do in the first place!" A blend with all components at equal intensity is an experimental artifact, and unlike natural mixtures. Hettinger added, "They would make sure it's thirty different things together at equal intensity, but this never happens in nature. You never have thirty things of equal intensity. You always have some that are higher than others, and some of them are much higher than others. So you'll have the dominance of single chemicals on odors or objects." These are essential considerations to olfactory psychophysics as proof of principle, as the case of olfactory white invites further investigation. Gilbert noted that the fact that differences in vapor pressure of individual ingredients may have less impact on the perceptual outcome (olfactory white) indicates the phenomenon might be robust. Margot concluded: "Conceptually, it's very nice to have done that."

The content of olfactory perception, as well as color perception, is determined by the sensory system that codes and computes physical features as neural signals. It matters, therefore, that the olfactory system involves coding principles different from the visual system (Chapters 6–8) since these principles determine the processes of perceptual categorization (Chapter 9). Besides, the difficulties of a linear stimulus-response model in olfaction do not end here.

Odor chemistry is fiendishly complex. Feature coding in olfaction deals with a multidimensional stimulus. About five thousand molecular parameters determine the causal behavior odorants in receptor binding. Compare this to the visual system, in which the low-dimensional parameter of wavelength defines the visible spectrum of color. Later chapters will detail how the systems differ in the coding affordance of their stimulus. For now, let's explore the very fact that the perceptual computation of the stimulus differs.

The Odor of Things

Throughout the twentieth century, chemists had hoped to predict smell from molecular structure. However, their driving interest was not the perception of odor, but the commercial production of fragrant synthetic materials. "I have to say there were a lot of chemists doing this stuff," Gilbert said. "And part of that was in the industry. People were developing new molecules and testing them to see if they were commercially reproducible. Those guys were trying to figure out . . . do hydroxyl groups make smells spicier? Is there some chemical coding? Which is a fool's errand, because the smell is in here," he pointed toward his brain. "The smell is not coded on any structure."

Fragrance chemistry continues research on structure-odor relationships, often without consideration of biology. There are pragmatic reasons for this strategy, Vosshall remarked: "They need to make perfume ingredients that will make fantastic perfumes, and they don't have a lot of interest in the basic science. They are in the business; they need to make money. And so they need to make ingredients, they don't have time for [biological details]."

But chemistry does not match perceptual space. Smith cautioned: "It's a fantasy of chemists that they have perceptual kinds once they have all that structure of odorants. It won't work. They can produce a little bit of force fitting, and that's as good as you can get. I don't think it's giving us any deep insight into the nature of odor perception, or odors as things which are intermediate between the molecules and the perceivers."

What defines similarity in structure-odor models anyway? Margot remarked: "If you think very thoroughly what makes two molecules similar or what makes them different, there're many ways to look at this. There is no absolute rule to find or to say two molecules are similar or dissimilar because on paper they may look very similar, but there is an abyss in between them, too. Like in the mirror images, the so-called enantiomers [mirror-imaged molecules], they look very similar, but in the end, they are very different."

Structure-odor relations in olfaction indeed are riddled with irregularities.[35] There are several examples of almost identical odorants that produce substantially different smells, such as isosteric molecules. Conversely, structurally distinct molecules can cause remarkably similar odors. An excellent case for this is musk, a prominent ingredient in many fragrances. Musks are of great variety, as you can see in Figure 3.2, where we find Tonalid, a tetralin musk; next to macrocyclic musks such as muscone; nitroaromatics such as musk ketone; and Helvetolide, an alicyclic musk. How does the nose know that they all smell musky?

Besides, monomolecular mixtures do not behave consistently in perception at all times. In a few notable cases, concentration changes alter the odor of an otherwise homogeneous solution. Take ethylamine ($CH_3CH_2NH_2$). Its quality, in concentrated form, is "ammoniacal" and, when diluted, "fishy." Likewise, diphenylmethane (($C_6H_5)_2CH_2$) smells like "orange" in concentrated form, and "geranium" when diluted.[36]

Keller, however, argued that the prominence of these examples reflects publication bias: "For 99.5 percent of odors, the perception of qualities doesn't change at all. Can I cite that? No—because nobody is boring and stupid enough actually to do that experiment.

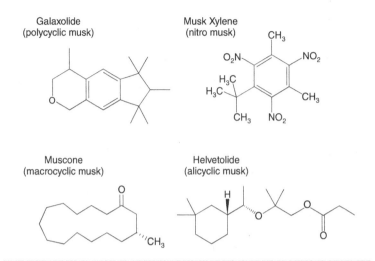

Figure 3-2 Examples of structurally diverse musk molecules.   Source: © Ann-Sophie Barwich.

But when they find one smell that changes quality with concentration, then they test it and write a paper about it, so I can cite that the quality changes with dilution. But I can't cite that it doesn't change. Everybody who works with odor knows that in more than 99 percent of the cases, it doesn't change. People who don't work with odors but know their literature think that for every odor with concentration the qualities change dramatically. That's an outsider case."

Regardless of the exceptionality of concentration changes, some cases do exist. The answer to such irregularities must be sought in the system that creates such effects. Acree explained that these cases, where odorants change their quality with concentration, are due to the presence of two different odor-active sites on one odorant (much like two different epitopes on one antigen). For example, 2-methyl iso-borneol ($C_{11}H_{20}O$) smells earthy at low concentrations and like camphor at high concentrations. If you look at the structure from one angle, it looks like camphor ($C_{10}H_{16}O$).

From a different angle, it looks like geosmin ($C_{12}H_{22}O$), and geosmin has an odor threshold one thousand times lower than camphor. If these molecules had the same threshold, no difference would be detected. Explanations about stability and variation in odor perception, therefore, require an understanding of the sensory mechanisms, such as binding behavior, not the isolated stimulus.

The puzzle of odor chemistry deepens with mixture perception. Ordinary olfactory experience does not involve single odorants, studied under controlled laboratory conditions, but compositions with multiple constituents. Almost every odor in your everyday life is what is called a blend. Blends are composed of several dozens, sometimes hundreds, of molecules. Coffee aroma contains about 655 volatile components, and 467 are present in tea. Strawberry has about 360 aromatic compounds, and tomato aroma 400. Even less-fragrant cases such as rice aroma involve 100 compounds; potato has 140.[37]

Blends act differently in contrast with monomolecular mixtures; they do not behave additively. Their quality is not merely a sum of their molecular parts. Margot added: "That means that a reaction that will work on two given substances under the same conditions may very well not work with two others. It's a question of developing intuition about getting to know how the molecule breathes." The disposition of an odorant changes sometimes unpredictably in different chemical solutions (as perfumers confirm).

Further, there are environmental conditions that determine stimulus composition. The nose regularly copes with unstable formations. Odor plumes in nature frequently alter composition. Visit a garden, and you experience olfactory differences based on the daily and annual rhythms of flowers. A rose in the morning varies a lot in its molecular components compared to a rose in the evening.

And yet we perceive its odor as rosy. Randy Reed noted: "Over many orders of magnitude, our ability to identify our perception of an odor is constant. We smell rose. It smells like a rose when it's ten feet away. It smells like a rose when it's two inches from your nose. The concentration of those compounds that you're supposedly detecting to give you the perception of rose, even if it was one pure compound, would be dramatically different." He explained the implications concerning the neural representation of odor: "The point is, what's active in the olfactory bulb is completely different at one concentration than it would be at another concentration. How do we filter all that out and still call that a rose? How do we get constancy of perception?"

Last, but not least, biology crushes the dream of direct chemical structure-odor rules. Consider the pig pheromone androstenone, the first mammalian pheromone identified. It is an excellent example of the explanatory primacy of biology in odor coding. Margot noted how the quality of androstenone depends on receptor sensitivity. "If you're highly sensitive, it smells like stale urine and elicits disgust reactions. If you have an average sensitivity, you describe it as having a wood-like odor or grass, or even a flower. And then there are people with very low sensitivity, they say it's faint, and it has no smell. A remarkable man who had studied that was Gary Beauchamp." Beauchamp also found that sensitivity to androstenone can change with age.[38]

One cause of variation is genetic. Margot referenced the (earlier-mentioned) 2007 paper by Keller, H. Zhuang, Vosshall, and Matsunami "reporting 7D4 as *a* receptor that responds prominently and selectively to androstenone and androstadienone. Unfortunately, they did not extend this to other structural analogs. That publication describes several genetic variants in the receptor, and you have two principal variants in the population that differ by three mutations. So if you're homozygous for the three mutations, you're more likely [smell-]blind [to androstenone]. If you're homozygous [meaning you have the same alleles for a gene on both chromosomes] for the wild or the original receptor you are likely osmic [able to smell]." These findings are only part of the story: "You can induce sensitivity to androstenone in [smell-]blind people—eventually, they will describe it as woody; or after a long time, they will even get this extreme sensitivity and say: That's urinous." At this point, Margot concluded, "this raises more questions than it provides answers."

The nose not only handles a highly irregular stimulus, but individual noses make different "scents" of the same stimulus. Isn't that astonishing? So the question is, how does the sensory system promote an overall stable perception? Behavioral studies tell that odor perception is remarkably constant, even in the face of a degrading stimulus (involving diminishing compositions of its odor plume).

Such odor constancy builds on learned pattern recognition, as Donald Wilson, a neuroscientist at NYU, confirmed with an experiment.[39] After training mice to recognize an odor mix, the removal of a component did not change their behavior toward the stimulus. However, the addition of a single element to the mix did. "We did this biometric study, where we tested mice on a ten-component mixture. We either removed components or added in a component that we were calling a contaminant. We looked at maybe a dozen mixtures in both cases. No matter what it was, the animals knew. They could learn very quickly that there's something added here. In the opposite case, they had real trouble noticing something was missing." Further studies confirmed the hypothesis of pattern recognition in olfaction.[40] This was music to Shepherd's ears: "It's patterns that are processed by the system!"

Ultimately, such examples of complex stimulus-response interaction demonstrate what smell is: odor perception is an interpretation of physical information by an organism. This interpretation, generating perceptual content, must be modeled via the biological mechanisms that code the stimulus and determine the patterns of recognition. Recognition involves both the physical stimulus and learned behavior.

Talking about odor objects as a stimulus representation is deceptive because it covers up the fundamental processes, biological and cognitive, that determine why odorants smell the way they do. Biology and psychology are the factors that resolve how physical information is organized into odor categories. You cannot blackbox the processes of perceptual coding. The danger, Gilbert reemphasized, is that "it's the same fallacy as expecting the odor quality to be encoded in the molecular structure somehow. Here, you're still expecting the odor quality to be encoded in a physical relationship that exists externally, and independent of the perceiver." Stimulus chemistry does not serve as a shortcut definition of odor objects as perceptual categories.

Odors as Elements of the Mind

Olfaction does not reside conveniently on either side of the conventional inner and outer divide by which philosophers discuss matters of the mind. This chapter revealed that explanations of perceptual content could not be shortened via reference to physical stimulus space. The perceptual experience of odors is irreducible to odor chemistry because it depends on the affordance of the physical stimulus as determined by the biological conditions of physical feature extraction, combination, and integration. Otherwise, Smith summarized, we would "confuse the vehicle and the content of representation."

Smells are elements of the mind. And the mind is better understood in terms of its dynamic processing than by analysis of mental snapshots. In this context, smells communicate multiple meanings in a range of perceptual activities, conscious and unconscious.

Odors turn into objects of cognition, for example, when we associate their perceptual content with semantic objects. Their conceptualization allows us to think and communicate our experience and directly compare it with other circumstances of perceptual experience. As objects of cognition, odors can appear in conscious experience and as part of multimodal perceptual images, even if we are not always aware of smells explicitly as olfactory sensations. Or they form objects of attention when we are aware of olfactory sensations without such impressions necessarily having concise conceptual content.

The manifold layers in olfactory perception reveal that its content cannot be explained without an understanding of sensory processing. It is profoundly misleading to speak of odors in isolation, as things without mental context. What gives smell meaning is bound to the processes in which its perception partakes. These processes define the olfactory sensations as targets of experience. Smells are an interpretation of physical information in the context of continuous operations, physiological and cognitive. In the end, the same stimulus can have various interpretations, and get processed into different odor images.

Consider the example of sulfurol (C_6H_9NOS). The master perfumer Christophe Laudamiel used sulfurol during a public symposium called "The Human Sense of Smell" in April 2017 at Columbia University.[41] Laudamiel distributed smelling strips soaked with sulfurol, encouraging the audience to smell and think about its odor. Reactions varied. Uncertainty filled the room as to what this smell was. It felt organic, a little sweaty, somehow sweet and fatty. It was not unpleasant; it was not very pleasant either. What was it? Laudamiel surprised the audience and showed a picture of warm milk. Of course, it was warm milk! A murmur went through the audience. Laudamiel changed the image to ham. You suddenly smelled ham! The same chemical, the same smelling strip, but your perception of it changed with the simple switch of an image.

Laudamiel changed the picture back and forth, and with these images also varied your experience of sulfurol. From ham to milk (the ham was gone!) and back. Keller reported: "It was like: change the image, change the odor!"

"It works for me too!" Laudamiel laughed. "I know the smell by heart. I know the pictures I'm going to show by heart. Every freaking single time, I look at the picture, I see the meat, I smell the ham. There's no control." He paused. "The brain learns very fast to do this. You saw that the public was untrained, had never smelled that molecule before. And, frankly, warm milk is not a common odor. People know it, but it's not like the smells they usually know."

This also works with blends. Laudamiel created another fragrance to similar effect: "One image then is a library with very dark wooden beams and red velvet. Here, you smell something spicy, the texture of the velvet, and all this. Then I show another picture with bookshelves. The wood is very sleek, very beige, clear woods, simple shades. All of a sudden, you don't 'see' the spiciness anymore. It's crazy. All of a sudden, you can smell the old paper inside the books." There are numerous versions of this kind of perceptual trick. In another one, you first smell a forest and fruits before Laudamiel shows a picture of a deer, "and you're like: Oh, my God. How come I missed that animal that's right there in front of my eyes? All of a sudden, you see a deer, which you hadn't seen."

An odor image is specified by the process in which it partakes. What does this imply for representational theories of perception—what makes an odor image an accurate representation of its source? The answer links to the mechanisms that create an odor image, physiological and psychological. Whether the categorization of information from the physical stimulus into a perceptual schema is accurate depends on the processes it serves.

Many odorants, like sulfurol, are ambiguous; they allow for multiple semantic attributions. The perceptual interpretation of a stimulus like sulfurol into a semantic object can involve various conceptualizations; here, Laudamiel used warm milk and ham. Additional odor images could have referred to beans. Representational "accuracy" thus depends on the affordance of a stimulus, as well as the conditions of its interpretation.

The idea of accuracy in perceptual representation, if connected to the idea of a designated universal percept, is a profoundly misleading philosophical idea. Say Laudamiel had shown you a picture of a rose, and you smelled rose. Then your perception would have been misleading—unless, of course, you have a very quirky mutation near some olfactory receptor genes (like with cilantro). Odorants can have multiple interpretations, yet not randomly so—their disposition grounds in the chemical features of the stimulus. Their perception, however, is based on the receptor repertoire of an individual. Alternatively, what if Laudamiel had shown you a picture of ham—and you smelled milk? The answer depends on the process you examine: do you test representational accuracy based on semantic congruency (cross-modal integration of an olfactory cue) or association (potential images linked to previous, alternative experiences of the same olfactory cue)?

The perception of smells is not merely a sensory abstraction from physical features, but an active interpretation under several measures (mental and biological). The real challenge is to identify the appropriate measures for modeling olfactory processing. These measures concern the behavioral value of smells (Chapter 4), as well as the various levels involved in the biological coding and computation of information from the stimulus (Chapters 5–8). Fixing the mental content of olfaction via stimulus chemistry fails to account for the dynamic experience of odor as a psychological activity. Or, in the words of Daniel Dennett: "This is like forgetting that the end product of apple trees is not apples—it's more apple trees."[42]

CHAPTER 4

How Behavior Senses Chemistry

THE AFFECTIVE NATURE OF SMELL

Smell allows us to select and respond to things differently. It alters our attitude toward things in the environment, and it alters our interactions with them. The sense of smell is an instrument for decision-making. Odors, quite literally, compel us to move. They repel and attract, they arouse or bring us to a halt. Bacon aroma turns your head; a whiff of feces makes you recoil; some fragrances, like lavender, are relaxing; others, like mint, refresh. Barry Smith emphasized the importance of this link between action and perception in theorizing about olfaction: "We better start looking that way. If we've learned anything in the last ten or fifteen years, in fact pretty much further back to Theodor Fontane, we realize that a theory of perception is not all about one-way traffic perceptions, being a mantra of passively taking in information from the environment. We look for perception-action links everywhere else, so why not with smell?"

The industry knows about "scent appeal." When you shop for shampoo, lotion, soap, or detergent—all things related to body and home—you select the one that smells good. Maybe you pick one that somehow smells right for a particular occasion (a date?). In the shopping aisle, it is easy to notice that the first thing people do when choosing a product is opening and sniffing it before making a decision. Shampoo might be effective in getting rid of dandruff no matter the smell, but the industry knows that it requires selling it in five or more different fragrances to market it to a broader population—a population that buys it simply because it smells nice.

Modern society is flooded with fragrant products. An entire enterprise is occupied with designing and selling them. In the United States alone, more than $28 billion is generated annually by scented products, which range from high-end perfume to scented trash bags.[1] "Believe it or not, those numbers are incredible," the master perfumer Harry Fremont revealed. "The fabric softener market is 1.2 billion, and those fragrances for scent booster products, which constitute a relatively new category, are already almost 500 million. It's amazing if you think about it."

The impact of smell on human behavior is real, not only when buying perfume or other commodities. The nose guides value judgments about our surroundings all the time. We actively choose and decide with our nose. Many underestimate the influence of smell on their minds. Not all decisions based on smell happen in awareness or evoke conceptual images. Smells can instruct behavior without residing at the forefront of consciousness.

Are you aware, for example, that you use your hands routinely as an extended tool for your nose to sample chemical information in your surroundings? Observe people for a while. You'll be surprised how often they touch their face with their hands to smell them. This unconscious habit caught the attention of Noam Sobel. His team documented how participants, unaware of the study's purpose, sniffed their hands after shaking hands with the experimenter.[2] (They also found that people, when sitting alone, spent about 22 percent of their time with at least one hand near their nose, displaying increased sniffing activity.) "Have you ever noticed yourself doing that?" Stuart Firestein prompted. "Yes. Absolutely," Don Wilson admitted. Firestein went on, "It really bothers me. Half of the time I'm even licking, which is really . . ." He ended abruptly, laughing.

The nose processes a lot of signals of which we are unconscious. That does not imply that these signals cannot reach the threshold of conscious experience. It just shows that focusing solely on the features we are consciously aware of fails to capture the fundamentals of perceptual behavior, including nonconceptual content. Our understanding of perception should mirror that fact. We won't understand what olfaction does if we restrict perceptual theorizing to propositional statements like "I smell oranges" or "This smells of roses." Excluding nonconscious and nonconceptual processes from perceptual analysis is to artificially amputate perceptual theorizing from the very conditions and elements that give rise to conscious experience.

The sense of smell communicates something about the quality of our surroundings that is not bound to result in concrete representations or conceptual objects. Perceptual recognition in olfaction operates by learned association. A range of feelings or emotive impressions can accompany these associations. The trade sector quickly realized and capitalized on that. Many hotel chains, for instance, employ perfumers to create a house scent, a pleasant background to sentience that customers come to associate with their brand.

Philosopher Claire Batty at the University of Kentucky surmised odors primarily as phenomenological sensations, or "feels," that are "free-floating" or "object-less."[3] Odors, she argued, appear as a manifestation of consciousness, or modification of conscious experience, that does not require representational accuracy of its causes. Something rings correct about this view. But it does not fully add up. Undoubtedly, odors provide a strongly qualitative layer that communicates how something appears to you, how it feels as an experience. Still, olfaction conveys more than free-floating feelings to the human mind. Olfactory experience conveys a concrete material experience. Such experience entices us to act directly in response to distinct environmental signals, and sometimes to identify their sources. Note that one of the last possessions family members of the deceased give away is that individual's clothes, as these textiles still possess the person's scent.

Affective content compels us to move and behave intentionally in response to what is around us. It matters in the evaluation of a stimulus what the perceived materials are, or how we categorize them. Jay Gottfried noted: "We put value judgments on things. A lot of these sensory systems serve a greater function than just simply identifying it. This is not to belittle the importance of identifying something, but ultimately these sensory stimuli in the environment modify one's behavior. The 'where' question can only be answered rationally by incorporating the 'what' question into it. In other words, you're not just going to wander, chasing after something that smells. It's something that is an olfactory signifier of an object you're trying to approach or avoid. In order to navigate towards a source, you need to be able to retain an internalized representation of what you're searching for, the 'what' question. It's folded together with the 'where' and how to access the source or avoid the source."

Affect constitutes an intentional state of the organism.[4] It communicates the value of something external. Instead of an objectless feel, it is directed at something in the world. It involves a judgment about that something. The measure of this value judgment is grounded in the perceiver; it depends on the perceiver's constitution and experience, and mental as well as physiological conditions (being pregnant, tired, hungover, or hungry). Affect is relational.

Relational does not mean arbitrary. Affect in olfaction results from an integration of multiple levels of physiological and psychological processing; in addition to context and exposure, it is defined by individual and cultural development. Olfaction mediates interaction with the world in various ways. Just what the nature of affect is, and what it does, is far from intuitive. Odor evaluation is not bound to specific physical features of the stimulus but is an interpretation of selected elements in light of the different processes taking place in the perceiving and behaving organism. What behavior does olfaction afford and inform? This question brings us to the heart of human olfaction in this chapter: behavioral responses to odors are contextual and learned. Such acquired connections can become powerful cues to evoke memories, sometimes accompanied by emotional tags. And so we get to olfactory memory next.

Odor Memory: What Proust Forgot to Mention

Smell, in popular imagination, is married to memory. Throughout history, odors were noted for their impact on personal memory and emotions. The French philosopher Jean-Jacques Rousseau is said to have declared (though never with citation), "Smell is the sense of memory and desire." Rousseau further found that "The smell is a sense intimately connected with imagination."[5] (This quotation is certified.) In the same vein, the American physician and poet Oliver Wendell Holmes Sr. later noted how "Memories, imagination, old sentiments, and associations are more readily reached through the sense of smell than through any other channel."[6] This idea lingered in the aesthetics of the perfumer Jean-Paul Guerlain, who noted: "Perfume is the most intense form of memory."[7]

The first thing that comes to mind about the intricate link of odor with memory is a famous episode by the French novelist Marcel Proust. In *Remembrance of Things Past*, Proust recalled an autobiographical incident that has become virtually synonymous

with public sentiment about the human sense of smell. Proust re-
lived how dipping a madeleine (a small French cake) in a cup of
tea brought back memories of his childhood:

> And soon, mechanically, dispirited after a dreary day with
> the prospect of a depressing morrow, I raised to my lips a
> spoonful of the tea in which I had soaked a morsel of the
> cake. No sooner had the warm liquid mixed with the crumbs
> touched my palate than a shudder ran through me and I
> stopped, intent upon the extraordinary thing that was hap-
> pening to me. An exquisite pleasure had invaded my senses,
> something isolated, detached, with no suggestion of its origin.
> And at once the vicissitudes of life had become indifferent
> to me, its disasters innocuous, its brevity illusory—this new
> sensation having had on me the effect which love has of
> filling me with a precious essence; or rather this essence
> was not in me it *was* me. I had ceased now to feel mediocre,
> contingent, mortal. Whence could it have come to me, this
> all-powerful joy?[8]

Proust recollected a vague but captivating feeling. At first, this
feeling was not connected to any concrete imagery. So he continued
to explore the psychological roots of his experience, separating
inner sensation from its current source:

> I was conscious that it was connected with the taste of tea and
> cake, but that it infinitely transcended those savours, could
> not, indeed, be of the same nature as theirs. Whence did it
> come? What did it signify? How could I seize upon and define
> it? I drink a second mouthful, in which I find nothing more
> than in the first, a third, which gives me rather less than the
> second. It is time to stop; the potion is losing its magic. It is
> plain that the object of my quest, the truth, lies not in the cup
> but in myself.

Smells yield an evocative force that can wholly overpower consciousness and enrapture human imagination. Sometimes "scentsations" break open an almost psychic window to the past. Psychologists, pertinently, termed such encounters the "Proust Effect."[9] Avery Gilbert appeared less dazzled by Proust's popularity. Renowned for lengthy prose, Proust reports an exhausting intellectual contemplation about the recollection of his memories. Gilbert noted in his book *What the Nose Knows*: "Proust's struggle with the soggy madeleine is distinctly not the way most people experience odor-evoked memory. For most of us, these recollections spring to mind easily. This is hard to square with his reputation as the sensual bard of scent." Besides, he added, "Here is another remarkable thing about the madeleine episode: it is utterly devoid of sensory description. Across four pages of text, Proust, that 'voluptuary of smell,' provides not a single adjective of smell or taste, not a word about the flavor of the cookie or tea."[10]

The puzzle with Proust is precisely that. The episode of the madeleine does not provide any insight into the actual quality of this experience. There is no answer to what defines the olfactory part in Proust's recollection. The experience of the madeleine is not about olfactory content. One could say that is not even about the smell. The aroma of the madeleine forms a placeholder for a psychological reconstruction of a memory—a memory that expresses a mental state of the perceiver without necessarily representing a concrete object in the world. Proust's experience is not about the madeleine he is eating, but a memory of the context in which he remembered having eaten the madeleine before.

Smells here act as gateways to all sorts of representations linked to the memory of an experience. While the olfactory experience is about something, it is not about the given, currently present, stimulus. This is a central clue to the behavioral role of smells. How does the brain make sense of odor memories to evoke past times or things? Are olfactory recollections more expressive than other sensory memories? The psychologist Rachel Herz, of Brown University, thought so: "Odor memories are more evocative. You're brought back to that original time and place. Say you remembered the time that you went to the movie theater with your grandfather to see a movie. The smell of popcorn could be the cue triggering that. If it's a visual trigger or verbal, tactile, or any other sensory form, you remember the content of the memory equivalently. But smell extends to what you felt back with your grandfather, the emotion that's connected to that memory is much stronger with the olfactory cue."

Theresa White added that intense emotional responses in autobiographic memory are not exclusive to olfaction. "Music will do that for you. You hear a piece of music you haven't heard for a long time, and it was important to you at some point in your life. It has a lot of emotion with it. Drags you back to that place. Odors can do that as well." She noted, "Obviously, they don't do it all the time. They can't. We encounter some odors so very often that they do acquire that semantic aspect, rather than the big episodic portion."

What makes odor memory mesmerizing is the feeling that it sometimes transports you, almost physically, to a different place or time. Smells evoke presence, an immediate physical embodiment. "There has to be a thing for philosophers and scientists here. Why we believe, when we remember a place or a person with a scent, it is so vivid and so real." Christophe Laudamiel proposed: "You are there because those molecules are there in your nose. And so your brain thinks that you are there. It's the same molecules. And then you have molecules that are different, and still, you can see yourself 'there.' It transports you in the place and in the exactitude of that place."

This tacit dimension of material presence seems to discern smell from other senses, Laudamiel continued: "When I remember a song, do I feel like I am inside this club in Berlin? No. If you were raised at church, you listen to a piece from church, do you see yourself inside a church? But if I smell frankincense, I see myself inside that monastery there. If I smelled that church scent, I would feel myself inside a church." Odor memory has an embodied presence. White emphasized that such evocative smell memories are rare compared to all the smells surrounding us daily. (Imagine being thrown back in mind each time you smell a familiar odor, ending up trapped in an endless loop of olfactory Groundhog Day!)

During the 1960s and 1970s, a range of studies promoted the idea that smell memory is unique and different from memories tied to other senses. A prominent researcher in this field was Trygg Engen.[11] The dominant hypothesis was that smell memory constituted an intimate connection through life. Once fixed, Engen advocated, the memory of smell was immutable. This idea aligned with the intellectual pull of Proust's madeleine, although critics called it a case of the "Marcel Proust Syndrome."

Gilbert remembered the rise and fall of Engen's idea with its widespread impact. "He was at Brown University. In the early '60s, he did those experiments that supposedly showed the immutability of smell memory. That Proust was right." But Engen's studies did not stand the test of time, Gilbert continued. "That stuff unraveled pretty quickly. Probably ten years later, it was like ... we don't believe this anymore. Smell memory, it decays, it changes over time, it can be tricked. But Engen was big."

Engen realized early on that any memory function of the senses—whether visual, auditory, or olfactory—depends on several factors, including the familiarity of the subject with the stimulus, her age, the use of memory aids, and so forth.

Bill Cain, at Yale and later University of California San Diego, was another prominent pioneer in odor memory. Cain showed that the memory of familiar and environmentally relevant odors, while not unchangeable, did have a higher retention rate in subsequent testing.[12] Their retention rate was notably higher than the retention of stimuli in other sensory modalities. But the precise mechanisms of odor memory, including its neural basis, continue to pose an issue for future inquiry.

Memory is not a uniform phenomenon. It constitutes an umbrella term for multiple cognitive and neural mechanisms. Talk about odor memory thus requires differentiation. On the one hand, there is *memory for odor* and, on the other, *memories associated with odor*.[13]

Explicit memory for odor, recalling odors as odors, is not a pronounced human ability. Most people, unless trained, struggle with remembering an odor; they cannot readily recall its label when presented even with a familiar smell (Chapter 3). You may also recognize familiar smells but cannot come up with their proper names. Engen termed this the "tip of the nose" phenomenon in 1977.[14] The notion relates to the "tip of the tongue" effect in which, accompanied by an air of familiarity, words or names cannot be retrieved. (Note that there is no comparable "tip of the eyes" effect.)

Acute short-term recognition and learning of smells do not automatically translate into long-term memorization or recall. Long-term memory of odors as odors requires effort and training. The functions of short-term and long-term odor memory in human behavior differ. Olfactory short-term recognition involves the detection of qualitative distinctions between stimuli, even minute differences if circumstances demand it. "The most important thing we have to do short-term with odors is to tell whether two things are the same or different," White clarified. "That's very short-term odor memory. It's important odor memory. If you're trying to figure out: are you making these things the same, or what was the last thing you just smelled and is this now the same sort of thing . . . those are useful things you can do with smells pretty easily."

What about memories associated with odors? Olfactory cues do not feature heavily as systematic mnemonic tools in human be-

havior. We don't use smells like visual symbols to access information. "Honestly, if I want to remember what to get at the grocery
store I don't go and sniff the milk and sniff the eggs," White
laughed. "It's what you have to be remembering, and the way that
you have to be remembering it, that makes smell important to
memory. When people say it's the best sense for memory, I think
what they're trying to indicate is its episodic nature. That feeling
of actually being back in the experience." Episodic memory of
odors builds on the salience of individually learned associations
between a smell and a memory for an organism.

Salience is not directly determined by the stimulus. Odors are
routinely memorized when a person has learned that the odors signify something beyond their own quality based on a contingent
occurrence. Herz concluded: "It does not make sense, evolution-
arily speaking, to come hardwired with any responses to the signals that are going to be meaningful for what is food and what is
predator—and olfaction is critical to that—but rather to learn on
the basis of experience, to learn very fast, to learn on the basis of
one experience, and that first experiences are really key." Meanwhile, first experiences are not fixed and immutable. The internal
conditions (physiology of the perceiver) and external circumstances
(features of the environment) of odor processing change regularly.
Odor experience, and the evaluation of its content, allow for re-
learning and recoding. Some odor memories last longer than
others, depending on their specific effects and initial exposure. The
feature that ultimately characterizes the link between odor and
memory is its inherent contextual encoding.

Odor memory is more than subjective recalls of experience.
Memory responses build on perceptual and cognitive processes of
a general nature, which are possible to study across individuals.
The specific effects might not be generalizable—however, the behavioral principle in question is. The proverbial smell of your
grandmother's kitchen, the perfume or cologne of your lover . . .

these are learned associations of variable physical stuff with familiar categories, which matter to the behavioral world of humans as social animals. You do not have to experience the particular smell of my father's garage, or Proust's madeleine, to know that it functions as a sensory reference to a recollection of childhood memories. Henceforth, the study of human olfaction must build on the mechanisms by which external signals are categorized and acquire behavioral meaning.

Smells as Moving Targets of Experience

Affect is not tightly scripted. It is not a predetermined response automatically triggered by stimulus structure. Except for a few notable cases, such as cadaverine (even that yields variations), most smells are markedly ambiguous in their hedonic assessments by humans. Ambiguous means that the same stimulus can have a different appeal in separate settings depending on the type of encounter. An interesting, unresolved issue is whether the system might process hedonically more "hard-wired" smells (like cadaverine) differently from more flexible ones (say, fatty smells). Notably, the effects of contextuality on hedonic evaluation applies not just to odorants that elicit little to moderate responses, but also odor associated with stronger physiological reactions, such as arousal and disgust. Think of body odor, which is considered unpleasant, even repulsive, when met via the armpit of a fellow passenger in the metro. In romantic encounters, however, it can have an arousing and positively stimulating effect.

Hedonicity is not engraved in stimulus structure. Hedonic responses follow contextualized processing through the sensory system. Herz offered a comparison: "Disgust is very similar to smell in that it has to be learned. What we perceive as disgusting is also contextualized, learned, totally dependent on all these different factors. It's all about the meaning you assign to the stimulus. The stimulus itself is basically agnostic. The meaningful way you apply the stimulus makes it good or bad. That is the critical feature in terms of the way that disgust and also smell work together." What defines context? A chief component is culture. People routinely learn to assign specific meanings to odors, which determine their effect and may vary across cultural contexts. Lemon, for example, is considered refreshing in central Europe because of its extensive use in cleaning products. Consumers came to associate these two notions, lemon and cleanliness, with each other. In countries where lemons grow in the stirring summer heat, accompanied by vast amounts of sweat and even more mosquitoes, the connotation with lemon scent may not always be clean. Meanwhile, its widespread association with cleaning products turned lemon into an unlikely ingredient in perfumery. It would not sell. Would you want to wear a fragrance that reminded everyone of a recently scrubbed bathroom?

Cultural associations with smells vary across times and places. Naturally, this also makes the design of a universally appealing perfume close to impossible. Fremont confirmed: "To make global fragrances accepted is very difficult. You need to have a very layered fragrance." Perfumes have changed significantly in style throughout history.[15] Nineteenth- and early twentieth-century fragrances for women carried heavy animalic, musky notes; this is something that would be considered obtrusive today in the cramped, crowded places of public transport and clean metropolitan lifestyles. Many classic perfumes contained rather scandalous odorants, Herz observed:

> Chanel No. 5 is in the scent category "floral aldehyde" and is composed of aldehydes, jasmine, rose, ylang ylang, iris, amber, and patchouli notes. But Chanel No. 5 also contains secretions from the perineal glands of the civet cat—secretions with a strong musky, fecal odor. The anal secretions from the

Himalayan civet cat, musk deer, and beaver (castoreum), and vomit from sperm whales (ambergris), have been historically used as perfume fixatives. Notably, with pressure from animal rights groups, the Chanel company claimed that as of 1998 natural civet was replaced with a synthetic substitute. Many of the most popular perfumes are laden with synthetic fecal notes, such as indole. Eternity (1988) by Calvin Klein is claimed to be one of the most indolic perfumes ever. . . . It is an interesting social observation that within the conservative palate of the 1950s and our modern obsession with ridding the body of its own body odor, the rise of perfumes that are redolent with fecal and bodily scents re-emerged. Yet, the presence of these funky, animalic notes was rarely advertised.[16]

Another excellent example for the divergence in cultural fashion is oud (or oudh), a raw material with a heavy scent that is extremely animalic and musky. Oud, a luxurious ingredient in perfumery, is often called "liquid gold." It is harvested from Southeast Asian agar (*aquilaria*) trees affected by a specific kind of mold. Opinion about the pleasantness and appeal of oud is notoriously divided. Oud is popular, especially in the Middle East, whereas people in Europe and North America find it obtrusive. Or consider the smell of wintergreen. If you are an American who grew up in the 1960s and 1970s, you likely associate this smell with gum or candy and probably consider it pleasant. A British person might resent the smell because it has been used in the United Kingdom in analgesic balms.

Next to culture, variations in the hedonic evaluation of smell ground in biology. Both short-term and long-term developments significantly shape how an organism is conditioned in its response to a stimulus. Thomas Hummel remarked that olfactory learning begins before birth: "What you like, what you don't like, may also be dependent, for instance, on what odors you were exposed to in utero. It already starts in the mother's womb that the baby is exposed to certain odors. That also shapes preferences. Excellent work has been done by a group in France. Benoist Schaal and colleagues showed that what the mothers eat during the last two weeks of pregnancy—if they eat anise flavored things—newborns actually show a preference for anise.[17] Newborns of women who do not have such exposure don't show that."

Olfactory learning continues throughout the lifespan of an organism. Mechanisms of biological and cognitive learning intertwine in this context, White added: "The role of learning in olfaction is huge. There's a good deal of evidence that the majority of the odors that we encountered, we've learned information about. I think not so much in terms of the categorization, but in terms of the pleasantness, certainly. You look at little children, and you see a real difference in how they interact with smells as opposed to adults."

"The hallmark of odor perception in humans is that you can change the valence of a smell," Gilbert agreed. Think about cocktail olives. "Nobody likes them immediately. Kids hate them. Once you learn to like a martini, you then perceive of a cocktail olive like: Oh, that's a nice fit." The opposite also happens, Gilbert continued. "You can come to hate a smell. Acquired aversions are very common in smell, and you can change your perception. So it's flexible. You can recode it. Those are higher cognitive processes." Hummel replied: "That goes on every day. Every day we change or can potentially alter our preferences." Flexibility is crucial for the sense of smell to fulfill its function: to help us make choices amid an ever-changing chemical environment and bodily states that require situation-dependent responses.

Smells also can be used to change the valence of other objects. Linda Bartoshuk explained: "Have you ever heard of evaluative conditioning? It's a branch of psychology that deals with the rules by which affect is transferred from valanced items to neutral items. Olfaction is its primary playing ground, but it applies to many different fields. You can take something that smells wonderful and put it in the presence of a neutral item, and it will wash onto the neutral item to become positive."

Affect in olfaction, Bartoshuk thinks, further links to motivation, and how motivation might be learned as well. "In the early years, people treated motivation as something that couldn't be learned. But it is learned. Olfaction is one of the beautiful examples, and taste, because we condition things: preferences and aversions. Many people in the history of learning thought this was silly. They wouldn't put room for it in their theories. So you have an interesting history developing: affect associated with motivation was ignored by the field. But when you begin to look at it, you have something like evaluative conditioning and Paul Rozin is the real hero here in my opinion." Rozin, at the University of Pennsylvania, was pivotal to advancing research on the evaluative function of odor in reactions such as contempt, anger, and disgust.[18] Bartoshuk emphasized the shift in understanding valence in olfaction: "You begin to ask yourself: What are the rules by which you can acquire affect?"

Another factor impacting hedonic judgment is intensity. The same stimulus under varying concentrations can change hedonic tone. Something might be pleasant in lower concentration while deeply unpleasant in higher concentration. Also, with prolonged exposure, the same odorous mixture (same concentration) can turn from pleasant to unpleasant. It is hard to pin affect to determinate stimulus features, and naïve to talk about predicting the molecular features of pleasure. Pleasantness is not reducible to molecular metrics. Affect is a biological, not a chemical, phenomenon, and biology does not work that way.

The biology of the olfactory system (Chapters 5–8) is geared to facilitate robust reactions to odor information. But information is encountered as highly contextualized cues (the same chemical in different settings affords different interpretations) and in relation to changeable physiological states of the perceiver (in response to which the stimulus is measured and made to make sense). Olfaction, drawing on its affective nature, facilitates informed choices about what gets into your proximity or what to avoid (such as food, people, toxins, or other items). Variation in sensitivity toward odors, under an organism's constitution, evolved for a good reason. Almost any odorant you encounter could kill you in higher concentrations, Max Mozell observed: "The theory is that most of the things you smell . . . if you drank them, they'd kill you. They are toxic."

Shifts in olfactory value and hedonics are responses handcrafted by your sensory system, not capricious subjective feels. Such shifts are causal reactions to external information tailored to your inner states, both physiological and psychological. Therefore, in perceptual analysis, we should foreground not the contingent expressions but the underlying mechanisms providing the rationale for behavioral generalizations.

The Epistemic Function of Olfactory Signals

Olfaction has an epistemic capacity precisely because of its inherent contextuality. To understand how the nose warrants judgment about the invisible reality of airborne molecules requires us to take a systems theoretical view, including the processes by which we learn to assign meaning to molecules.

Medicine is an excellent example to demonstrate the epistemic function of smell in human behavior. Olfactory cues have a long history and continue to play a vital role in medical practice. Medicine is a smelly business. Diseases emanate fumes of decay and infection, and—before pills—remedies came in distinct aromas. The medicinal powers of plants, especially, were linked to their healing powers (recall Linnaeus, discussed in Chapter 1).

Old disease conceptions, like the miasma theory propagated centuries ago, saw odors acting as infection carriers. Distinct smells accompanied particular diseases. Although incorrect, the miasma theory resulted in the enforcement of several odor-related social policies to prevent the spread of diseases. This idea stuck; air purification remained a prominent concern in the eighteenth and nineteenth centuries.[19] As an example, Central Park in New York was designed as the city's lung to counteract the harmful fumes of industrialization.[20]

Body odor still offers plenty of indications about a person's health.[21] Some odors point toward metabolic and dermatological dysfunctions; they suggest enzyme deficiencies or infections. Not all diagnostically relevant odors smell foul. The "maple syrup urine disease" is named after the peculiar smell (sweetish maple syrup) exhibited by the urine of affected infants. If not treated within a short period, the outcomes of this disease are irreversible neurological damage, seizures, coma, and finally death. Another disease of genetic origins results in a fishy-smelling body odor ("fish odor syndrome," or trimethylaminuria). The point of these examples is to show how smells perform a certain epistemic function in human behavior. Olfactory cues prove reliable tools to form preclinical hypotheses about an underlying condition. Accurate judgment can quite literally involve life or death.

Beyond general practice, smell plays a role in basic research, specifically in studies of cancer and neurological disorders such as Alzheimer disease and Parkinson's. Recent research on skin cancer and other carcinomas started looking for alterations in the chemical composition of body odor. Can specific changes in the body odor profile of a person signify a hidden disease? The use of medical detection dogs tells us as much. These dogs are trained to single out tissue samples infected with cancerous cells or detect lung cancer in the breath of a person. Long derided at as quackery, a few breakthrough studies put a stopper on premature dismissal.[22]

Dogs are known for their excellent noses. But humans, too, can pick out disease-specific odor cues. There is the curious case of a British nurse, Joy Milne, whose husband died of Parkinson's and who discovered that she could sniff out the illness from the sweaty shirts of patients:

> About a decade before her consultant anaesthetist husband was diagnosed, Joy noticed she could detect an unusual musky smell. Joy said: "We had a very tumultuous period, when he was about 34 or 35, where I kept saying to him, you've not showered. You've not brushed your teeth properly'. It was a new smell—I didn't know what it was. I kept on saying to him, and he became quite upset about it. So I just had to be quiet." The retired nurse only linked the odour to the disease after meeting people with the same distinctive smell at a Parkinson's UK support group. She told scientists at a conference, and subsequent tests carried out by Edinburgh University's Dr Tilo Kunath confirmed her ability.[23]

Scientists have begun collaborating with supersmeller Milne to determine the chemicals she's sniffing out, developing a first diagnostic test for Parkinson's based on body odor.[24] Training and selective attention may turn the nose into a potent and highly accurate instrument.

Knowing what the biological nose knows might aid in developing artificial intelligence, too. Andreas Mershin, a biophysicist with interdisciplinary vision at MIT, uses medical detection dogs as a model for bioelectronic noses in cancer detection. Mershin went to work with the organization of Medical Detection Dogs in the United Kingdom to test this idea after reading the book *Daisy's Gift* by Claire Guest.[25] For Mershin, one thing was clear. We need to rethink the basis for bioelectronic noses to make them a success. Machines should be modeled more closely to the organic systems that evolved to do the tasks we want to teach the machines. We may even learn about the biological processes along the way.

Smelling itself is a tool for diagnosis.[26] Decline in the ability to recognize odors can signify the onset of severe neurodegenerative disorders, such as Alzheimer's disease, Parkinson's, or Lewy bodies. Hummel said: "The first work has been done in the '70s, and then a major player was Dr. Doty [in the '80s]. He and others showed that when you have a neurodegenerative disorder, your sense of smell is mostly compromised in very many cases." Richard Doty was the inventor of the first clinical smell test in 1984, the University of Pennsylvania Smell Identification Kit (UPSIT).[27] He explained: "In things like Parkinson's, the smell loss occurs years before the onset of the classic clinical motor symptoms. We think that the smell system is affected very early in these disease processes. Indeed, some agents from the environment, for example, get into the brain through the smell system. They could initiate some of the pathologies of Alzheimer's and Parkinson's within the olfactory bulbs."

Next to Doty, Hummel himself had been leading research on the clinical effects of smell loss and its connection to cognitive decline.[28] Hummel was involved in the development of the Sniffin' Sticks, another test kit, in 1997.[29] Unlike the UPSIT, the Sniffin' Sticks separate three different olfactory functions: threshold, discrimination, and identification.[30]

Hummel noted: "You see the same decline of olfactory function in Parkinson's disease as you see it in Alzheimer's disease. From the outside, it looks all the same. Of course, the symptoms are different in patients with Parkinson and with Alzheimer's; also, in Huntington's disease and other neurological disorders. Just doing a smell test tells you this person has no good sense of smell, has no sense of smell, or has a normal sense of smell. It doesn't tell you what disease it is. But a smell test can be clinically very effective! For instance, a patient comes to us, and the patient has signs of

Parkinson's disease. This patient has a normal sense of smell. Then we would send him out again to redo the diagnostics. Because probably, this person does not have Parkinson's disease, but a different form of neurological disorder."

Doty and Hummel work hard to demonstrate that olfaction plays a vital part in clinical diagnosis. Doty added: "It's still challenging. Despite the interest in certain groups, the medical community still has not fully incorporated the idea they should be doing smell testing in neurology."

The nose is an exact instrument from which we can learn a lot about the world and ourselves. Olfaction has a truth function; it grants us epistemic access to detect real features in the world, when we are trained to do so and when it matters behaviorally. In one way or another, the nose plays a significant role in all things that shape human life: danger, food, pleasure . . . and sex.

Love, Sweat, and Tears

Let's finally talk about the obvious topic: sex. We like to marvel at the belief that people court each other because they have discovered a deep romantic connection. More commonly, it is a matter of physical attraction. Part of that attraction is olfactory. Other animals pick out partners by scent. Is that also true for humans?

Comparing mice and men, Gilbert wanted to know. During his postdoctoral years at the Monell Chemical Senses Center in Philadelphia, and working with Kunio Yamazaki, Gilbert noticed something peculiar. Mice smell the difference between two genetically identical strains of mice that differ only in major histocompatibility complex (MHC). And they prefer to mate with the opposite strain. "I got interested in that and thought—can we smell what the mice are smelling?" Gilbert recalled. "That led to my first human experiment, which was putting mice in little sandwich boxes, and having people smell and say if they were the same or different. It was sort of a man bites dog story," he laughed. "I also collected vials of mouse urine and little vials of mouse turds and had people smell those. And by gosh, as a team, a group of people could distinguish between the two types of mice. Single gene difference, urine, feces, or whole body, you could tell the difference! This is still one of my most cited papers.[31] It's just this nutty experiment. But we can do what the mice can do, olfactorily." Humans can do the same thing. We sniff our way to romance. Women were documented to favor the sweat of potential partners with a dissimilar type of human leucocyte antigen (HLA)—the equivalent of the MHC in mice.[32] Moreover, preference changes with hormone changes: women on contraceptives (the pill) prefer partners with a similar HLA type. Unfortunately, the love of your life could lose some sex appeal with changes in contraceptive routine.

There is more to social life than sex. Odors participate in other forms of human interaction. "What I find personally very interesting is the social odors," Hummel revealed. "How we communicate through odors. There have been a couple of very nice examples, like fear perception through odors, or the perception of imminent disease, or also that female tears are substances that have changed male libido."[33] Other examples are the learned preference of infants for their mother's smell during breastfeeding.[34]

A word of warning, though: It is easy to fall prey to spurious correlations in this context. Maybe you have heard that olfactory cues influence menstrual synchronicity in women, a hypothesis that has been debunked, or is at least insufficiently supported.[35] The difficulty with this sort of study, Hummel mentioned, is that "you need to have well-controlled experiments. It is not easy to study. You need to have large groups of people, and it's a difficult experiment." Human behavior is complex and instructed by a variety of dynamic factors. Too many factors shape the outcome; not all of these influences are known in advance. Decisive studies, therefore, are hard to come by. The latest discussion on reproducibility in psychology and the social sciences has highlighted this fact. That should not stop further exploration.

To date, the real extent to which odors shape human interaction is unknown. A variety of interesting, underexplored research topics are up for grabs. (This is where the mind of Sobel has set sail. Sobel's team currently conducts a broader study on "the smell-based social network," in which they test whether people with "a similar sense of smell have good relationships.")[36] What is certain is that odor communicates essential social cues in human life.

A Brief History of Pheromones

Talk of sexy smells conjures up another term of widespread interest: pheromones. Are they a myth, a marketing tool, or reality? Gilbert cautioned: "I think the pheromone concept has been stretched so far to fit human experience that it does not work anymore. Do I believe that we do kin recognition, emotional detection with smell? Yes. Mate selection, MHC, HLA. Chemosignaling is happening all the time. But the pheromone concept, I don't even use the word anymore."

Gilbert reminded us of a scientific concept with a similar fate: "Remember instinct? After Darwin, there was a period where there was this huge rise in the use of the term. Everybody's explaining everything with instincts. There was the instinct for maternal behavior. There was an instinct for self-preservation. Instinct for emptying your ashtray when it was full. It got ridiculous. And then people stopped using it. It lost explanatory value. I think that happened with pheromones."

Pheromones are a subset of smells. Pheromones refer to molecular signals that elicit species-specific behaviors or physiological responses in a conspecific. It was thought at one time that, in vertebrates, pheromones had their own exclusive pathway: the chemoreceptive vomeronasal organ (VNO) separated from the olfactory epithelium. The accessory olfactory system processes signals from the VNO with close connections to the amygdala and hypothalamus. Of course, it is not as simple as that, the Oxford zoologist Tristram Wyatt explained: "So you know all about the second nose, the VNO, the vomeronasal organ. Most traditional neuroscientists were aware that the VNO and the MOB and MOS—the main olfactory bulb, the main olfactory system—all the tracts come together in the amygdala. At the periphery, at the receptor level, you have two different organs, different families of receptor proteins. But higher up in the brain, the neuroscientists knew that all these inputs converged."

The convergence between the accessory and the main olfactory system was debated, Wyatt continued: "Molecular biologists in the late 1990s, people like Catherine Dulac at Harvard, argued that all mouse pheromones would be detected by the VNO, and anything detected by the VNO would be a pheromone. Lots of papers in the late 1990s, early 2000s conflated pheromones with the VNO. However, in 2006, Catherine Dulac's and Linda Buck's labs independently showed that indeed the signals from the two olfactory systems are integrated in the amygdala and other parts of the brain and there were top-down interactions too.[37] Meanwhile, some mammal pheromones had been shown to be detected by the main olfactory epithelium and not by the VNO."

There is disagreement over what pheromones are. The origin of the term offers help. It started with insects, Wyatt explained: "In the 1880s, the first state entomologist in New York State observed male moths flying to a virgin female and reasoned that if only we could identify and synthesize these powerful molecules, then we could use them in pest control." The word "pheromone" is recent. It was coined in 1959 by two biologists, Peter Karlsson and Martin Lüscher.[38] They were prompted by their chemist colleague Adolf Butenandt's identification of the first chemical structure of a pheromone: the female sex pheromone of an insect, the silkmoth. Wyatt penned a *Nature* piece for the fiftieth anniversary of the pheromone concept.[39] Afterward, he received an email "from a Greek professor who had been a PhD student in the lab in Germany when the word was coined." As a native Greek, the student protested that the word should be *pheroRmon*. "But for phonetic reasons the r was disposed of." That's why we have pheromone. It's bad Greek, but a word you can say. Wyatt pointed out that "pheromone" also sounds like "hormone," which fits its signal role. Yet unlike hormones, pheromones operate *between* bodies rather than within one. The new word immediately replaced an earlier word, "ectohormone."

When did pheromones enter research on mammals? "I've mentioned Karlson and Lüscher because their *Nature* article proposing the new word anticipated pheromones would be found in vertebrates, not just invertebrates."

Tides turned quickly against the idea of mammalian pheromones. "A few more years into the late '70s, the people working on mammals started to think it's going to be very hard to identify any pheromones at all in mammals and, perhaps, we won't find any at all. We won't have this kind of instinctive and innate behavior in mammals, unlike insects, so it is pointless to look. They published a manifesto, a bit like a group of artists, where they say mammals don't have pheromones.[40] Pheromones are a concept applicable only to insects. And they kind of rewrite history."

Wyatt added, however, "in the 1980s, you also have people saying that mammals *do* have pheromones. One was a chemist called Milos Novotny who published a series of papers on mouse pheromones.[41] In 2003, Benoist Schaal in Dijon, France, found the rabbit mammary pheromone, which prompts suckling by rabbit pups.[42] Since then many small- and large-molecule pheromones have been found in the mouse and other mammals [for example, by Stephen Liberles at Harvard].[43] Many of the smaller molecules are detected by the MOE [main olfactory epithelium], whereas protein pheromones, such as the mouse pheromone darcin, are detected by the VNO."

"Perhaps we need to separate the term pheromone from the individual odors each mammal gives off," Wyatt offered as a solution.[44] "I introduced the idea of *signature mixtures*. Whereas pheromones are the same in every male (though dominant males may produce larger quantities), each male also has its own individual odor profile, its signature mixture, which other mice learn so they can recognize it. For a scientist, if you're trying to identify a pheromone, then you're looking for molecules that are the same in every male." Pheromones differ from signature mixtures. "With these you're looking at a very different concept, of how animals tell each other apart, how they recognize kin and nonkin, and how they recognize their partner or detect a stranger—all that is about difference . . . so the researcher is looking for the differences in odor molecules given off by every individual. Pheromones are superimposed on this individual profile."

Thinking about pheromones has something in common with research on the main olfactory system. In each system, behavioral flexibility and molecular complexity make clear-cut conceptual distinctions difficult in the study of organismal responses. As chemosensory signals, pheromones do not differ from smells in terms of molecular structure. The difference lies in how an organism or species (as a collective group of organisms with shared evolutionary origins and history) responds to a stimulus.

Steven Munger, at the University of Florida, concluded that one could think of "odors in two categories. One category includes compounds (such as pheromones, kairomones [mediating interspecies communication, chemical signals benefitting the receiver such as parasites], and allomones [interspecies signals benefitting the releaser not the receiver]) for which a particular biological meaning has evolved within or between species and which elicits

or facilitates an innate response in the recipient. The other includes compounds that are going to be part of an environmental pattern that an animal will learn to associate with other sensory stimuli or contexts. In the latter case, the meaning is learned. Sometimes, a compound can fit in both categories."

Biology, not chemistry, determines whether a molecule gets processed as a pheromone or a "regular" odor. Stipulation of what makes certain features more likely to elicit a specific reaction, therefore, depends on what the system is wired to pick up and how. What is needed is a systems theoretical approach accounting for the different levels by which behavior must be measured, analyzed, and compared.

CHAPTER 5

On Air

FROM THE NOSE TO THE BRAIN

Odor perception facilitates spatial behavior. What is the point of detecting behaviorally important signals, like smoke or bacon aroma, if you cannot figure out from which direction they come? The origins of an olfactory cue may not be visible or immediate, but that does not preclude smell from aiding orientation and navigation in space.

"Smell is a distance sense," Jay Gottfried remarked. Its behavioral function in spatial interaction, such as foraging, indicates crucial principles that also define its computational processing. "By the very virtue of it being a distance sense, it is a predictive sense." Predictive in this context means that our brain anticipates stimulus regularities, based on prior experience, which allows orienting behavior in time and adjusting motor responses accordingly. "Because if you're smelling things from afar," Gottfried continued, "and if they're important to the smeller, then the point is not simply to marvel at the interesting smell but to approach it and figure out how to navigate towards that smell . . . like if it's food, or a mate, or home, or whatever. Or if it's something aversive—a predator, a fire, whatever the case. A peak function is to bring the smeller closer to things that it needs to obtain and further away from things it needs to avoid." Smells move us, emotionally and affectively. But how do we move physically toward their sources? Spatial orientation via smelling is not immediately intuitive. Despite its capacity to instruct spatial behavior, the sense of smell does not present its content as being similarly spatial in comparison with vision, which conveys discrete spatial qualities of distal objects, including their size, shape, and orientation.[1] Visual objects possess an extension; they communicate a location, and they display movements or direction. Additionally, visual objects are discrete in that they possess more or less definite boundaries; they routinely have a distinct beginning and end.

But what are the spatial qualities of odor? Talking about the orientation of a smell sounds odd. A visual object can face you, but it is hard to imagine how it would be to have a smell facing you in the same sense. Other forms of spatial descriptors, common to visual experience, are similarly difficult to apply to fragrance and aroma. When savoring wine, we do not describe the blackberry note as sitting on the left or right side of the tobacco note. This lack of spatial dimensionality is not an issue of fuzzy borders between multiple odor notes in a complex mixture. Olfactory experts excel at picking out individual notes. When they describe a fragrant note as sitting "beneath another note" in the evaluation of aromatic artifacts—like in wine, perfume, or whiskey—such an expression is a metaphorical description of phenomenological figure-ground segregation. (Figure-ground segregation allows the perceiver to pick out specific information against a noisy background.)

The nose does not convey spatial features of perceptual objects like the eyes do. But the nose allows us to behave spatially in relation to these objects. A comparison with other distal senses, vision and audition, exposes the causal basis of this capacity. Overall, the senses grant us knowledge of what the seventeenth-century philosopher John Locke termed primary and secondary qualities.[2] Primary qualities are mind-independent, primarily spatial properties of objects that include solidity, figure, extension, motion, and rest. Secondary qualities are mind-dependent properties, meaning they are not strictly properties of objects but a creation of the mind that perceives these objects.

Sounds are secondary qualities; they depend on the perceivers and their sensory apparatus. As an example, ultrasonic frequencies have no perceptible quality for humans because our auditory system is not attuned to them. But bats use these frequencies to echolocate and hear something humans cannot detect. It is impossible to answer "What is it like to be a bat?," a question posed by the philosopher Thomas Nagel in 1974.[3] This is simply because, without the appropriate receptive system, the physical stimulus does not convey a secondary quality. The same philosophical argument holds for colors, odors, and tastes. Just imagine, the mantis shrimp has fifteen different color receptors. Whatever this creature sees, it is unlike the human world we perceive.

As a secondary quality, color is not spatial per se. There is nothing inherently spatial about color as a color. Spatiality enters visual perception when color is perceived as an integrated feature in the visual representation of objects and scenes. Color then communicates spatial distinctions. Painters know well that color contributes to the perception of the distance of a visual object, even to its texture or shape. A walk through the galleries of the Metropolitan Museum of Art yields profound insight into the principles of the visual system (more than philosophical treatises do), as the neuroscientist Margaret Livingstone illustrated in *Vision and Art: The Biology of Seeing*.[4] Visual objects result from an integration of separate kinds of feature processing. Here, vision builds on two feature processing pathways: color coding and edge detection. Edge detection underpins the central coding of spatial dimensions.

Recent research, however, has called into question a strict division between edge detection and color vision in visual object formation: color coding is not solely processing color, but also resolving spatial features like edges. Neurophysiological studies have revealed considerable overlap in both pathways, even in the periphery. Understanding of vision is less settled than we think.[5]

Olfaction, too, involves more than one dimension. Next to quality ("rosy") and hedonicity ("pleasantness")—detailed in Chapters 3 and 4—there also is intensity. The neural processing of intensity is little understood to date.[6] Intensity, based on concentration coding, is not coextensive with the principles of quality coding. Intensity affords spatial behavior in olfaction.

Joel Mainland highlighted intensity as one of the critical features of olfaction, noting the difficulty and current neglect of its study. "The problem with intensity was that it [is] difficult to get an animal to report intensity, and make sure that the mice are reporting intensity and not quality shifts, or some other correlate. Yevgeniy [Sirotin] did this really nice experiment, where he showed how to do it, that you get these smooth transitions.[7] It's not like you abruptly transition from pleasant to unpleasant, but rather have a smooth concentration gradient that defines how the rat responds. So he did work where he says: Here's the intensity. Then I give you a higher concentration, but adapt you so you perceive it as the same intensity as the lower concentration. The physical stimulus is different, but your perceived intensity is the same. What's the neural correlate of the perceived intensity?"

Mainland explained what is at stake: "Once you find concentration A of odor A and concentration B of odor B where they have the same perceived intensity, you can give two totally different stimuli activating totally different sets of receptors to the animal and figure out, how is it encoding intensity? It's difficult to do these experiments until you work out some very basic principles of what perceptual information is being passed, keep it tied down to some behavior in the end, and then follow it through the system." Intensity coding presents a mystery. Its solution promises a better understanding of how animals navigate their environment with olfactory cues.

Stimulus Space

How does odor perception hook up with the motor system? At the level of the physical stimulus, odorants are spatial objects. Molecules are made of solid matter; they possess extension in space. Odorants are volatile chemicals and move around to occupy different positions in the environment. In principle, the olfactory stimulus affords an organism the potential to discern its location in the environment (where does this bacon smell come from?), as well as its directionality in relation to the perceiver (it seems to be coming from my left!). This unites vision, audition, and olfaction in their function as distal senses; they allow organisms to orient and move in relation to objects in space.

The distal senses also differ. It is not a feature intrinsic to the stimulus that facilitates olfactory navigation in humans and other animals. It is how the stimulus is detected by the system: what regularities can be detected, and how these regularities are represented as a sensory signal. Here, the olfactory stimulus differs from the visual and auditory in one respect: predictability. Odorants are highly volatile objects, distributed irregularly and moving continually around in the environment. Unlike photonic surface reflections in vision, the movements of odorants in the atmosphere are hard to predict or control. Gottfried described how this resonates with differences in the makeup of the sensory system. With a predictable stimulus, it is possible to have vision wired up as a topographic system where "the activation in our retina can always be linked to that spot in the primary visual cortex. But with odor, depending on the wind currents, it might end up coming from a different place." Spatial perception depends on how the stimulus hooks up with the sensory system.

The olfactory stimulus, in its environmental occurrence, has no regular or predictable spatial correlation to its local detection in the epithelium or neural projections. In *Consciousness Explained,* the philosopher Daniel Dennett highlighted this as the reason why olfaction has a lower spatial resolution than human vision:

> We may be able to sense the presence in a room of a thin trail of formaldehyde molecules, but if we do, we don't smell that there is a threadlike trail, or a region with some smellably individual and particular molecules floating in it; the whole room, or at least the whole corner of the room, will seem suffused by the smell. There is no mystery about why this should be so: molecules wander more or less at random into our nasal passages, and their arrival at specific points on the epithelium provides scant information about where they came from in the world, unlike the photons that stream in optically straight lines through the pinhole iris, landing at a retinal address that maps geometrically onto an external source or source path. If the resolution of our vision were as poor as the resolution of our olfaction, when a bird flew overhead the sky would go all birdish for us for a while.[8]

That is what distinguishes vision, audition, and olfaction in their performance as distal senses. Odorants, unlike the stimulus in vision and audition, are unpredictable regarding their precise physical trajectory in space, their placement and movement. This has consequences for the kind of interaction the stimulus affords the system, and how the system evolved to best deal with such stimulus. Consider all the possible ways in which the various compounds of an odor plume could be thrown together in the air, randomly shifting and changing their position every millisecond. It is simply beyond the brain's capacity (or even need) to calculate the precise arrangement of individual odorants in space. In a mixture coding for the quality of pineapple, for example, it does not matter whether the compound of allyl hexanoate ($C_9H_{16}O_2$) is floating "in front" or "to the right" of another component, say, ethyl maltol ($C_7H_8O_3$).

What matters to sensory processing in the olfactory system is not how molecules are distributed in physical space (as a consequence of airflow turbulence), but how these molecules interact with the epithelium—where the components of an odor plume are separated by temporal interaction. To spatially track odors requires a reasonable enough estimate of where a cloud of molecules is coming from, and maybe how far away its source is. Research on insect olfaction has suggested that olfactory navigation seems to build on the processing the temporal structure of odor plumes. Some of the primary researchers in this area are John Hildebrandt, Ring T. Cardé, and Michael Dickinson.[9]

Spatial dimensions in odor perception are not intrinsic to the stimulus; rather, this information is formed in relation to the perceiving system. Olfaction is a contact sense of molecules interacting directly with the epithelium. Olfactory cues signal things relevant for action and also act to signify distal elements (such as smoke). In effect, smells are qualities of something caused by volatiles emanating from the surface of things. These material origins of odor are localizable in space. Air turbulences exacerbate the spatial estimate of odor sources, however, since there is no continuous gradient. Massimo Vergassola et al. showed that infotaxis with olfactory cues thus requires a zigzagging strategy for odor plumes that are encountered in random and disconnected patches.[10]

Smell has a clear exteroceptive, outward-oriented dimension: odors afford movement toward or away from something. And so olfaction is a spatial sense if understood as signaling external targets of behavior and in relation to an organism. But its spatial dimension must be understood via the sensory system, not the stimulus.

Enactive or Embodied Space

Central to olfaction is sniffing. Sniffing is not a monotonous or automatic behavior. Variations in sniff speed, strength, and patterns determine how, and even which, molecules reach the nasal epithelium to interact with the mucus. Max Mozell emphasized the complexities of calculating airflow dynamics and flow currents in the nose: "We know there is specificity in the receptors themselves. But that doesn't tell us anything about how the molecules move over to the receptor field." You have to account for the number of molecules, volume, and time—and "each of those, in different combinations. Concentration becomes the number of molecules divided by time, and flow rate becomes volume by time. It's very difficult to determine from the response which one of these variables is more important. All three of the basic variables play a role depending upon how they're combined."

Sniffing influences what information reaches the brain and when. The volume, duration, speed, and strength of sniffs are measurable parameters that modify the perception of odors by adjusting how volatile molecules interact with the nasal epithelium. To reach detection threshold, lower concentrations of odorants require stronger sniffs compared to considerably higher levels of concentration.[11] There even is suggestive evidence that sniffing might elicit something comparable to odor percepts in the absence of a stimulus.[12] Sniffing entails more than the mechanical means to transport smelly molecules to the epithelium. It is a formative element in the generation of perceptual content.

In humans and other mammals, the olfactory system evolved to aid olfactory navigation through a nasal cycle, an involuntary mechanism in which respiration speed between nostrils alternates. "What varies is swelling of the nasal epithelium, resulting in changes in airway patency, and thereby airspeed," Avery Gilbert clarified. Although you are unaware of it, your two nostrils breathe at different speeds! One side of your nose is always slightly clogged up. So one nostril is a bit slower in drawing in the air than the other. This is not a permanent condition; the two nostrils take turns. Conventional wisdom has it that the nasal cycle shifts about every two and a half hours.[13] The implicit hypothesis of rhythmicity, however, did not hold up to scrutiny, lacking statistical evidence of rhythmicity. Two studies by Gilbert found that "nostrils alternate in patency, but on an irregular," meaning nonrhythmic, basis.

Rhythmically or not, by varying airflow speed the nose can detect a greater variety of odorants. Depending on weight and size, some odorants travel faster and interact more quickly than others with the receptors in the nasal epithelium. Additionally, odorants differ in the speed of their binding behavior. Smelling in two different airflow rates allows the nose to cast the stimulus net wider. It also lets your nose discern differences in the directionality of an odor source.

A look at dogs shows how this works. Dogs, picking up a scent, follow a trail. They do not follow a trail linearly or directly, however. They circle and sniff around that scented trail to hone in on its direction. They "see" where a scented trail leads as little as humans do. Dogs discover and follow the course of an odor plume by moving around its edges. This principle applies to all animals, including moths, fish, and humans; the neurobiologist Tom Finger at the University of Colorado explained: "You can get distance from how sharp an odor plume edge is. As odor plumes come off sources, the molecular edges of the plume are quite sharp. The source is putting out a fixed concentration of odor. As you go away from the source, that edge blurs because you get diffusion across the boundary. An animal can get an idea of a distance by how sharp the odor plume edges are. That I think is another dimension [in the perception of odor] that people generally don't appreciate. The sharpness of transitions from being 'in the odor plume' to 'out of the odor plume' is very informative about the distance from the source."

That is also how humans navigate with their noses. Gilbert mentioned a 2015 study by the psychologist Lucia Jacobs at Berkeley.[14] In this study, he said, "She had people blindfolded and with earphones, put them in a room, and there was a smell in one place, and she spun them around, and they had to orient themselves with respect to a single source smell, and they could do it!"

Contrary to popular belief, humans aren't worlds apart from dogs in their ability to track scents. This was shown with a 2007 study by Noam Sobel and graduate researcher Jess Porter modeling the human ability to track scents after that of dogs.[15] They had thirty-two hungry Berkeley undergraduates follow a chocolate-scented trail. To make sure the students were informed only by the smell, not other cues, they artificially deprived their subjects of other sensory input (using blindfolds, gloves, knee pads, etcetera). The students' actions in tracking scents were compared to the behavior of a dog sniffing out a pheasant's trail. The students showed tracking behavior similar to that of a dog, following the path by circling its track. Plus, with training, the students became faster and better at this task!

Claiming that the human nose is equal to that of a dog, however, might be premature. Alexandra Horowitz is a cognitive scientist studying dogs at Barnard.[16] She intervened: "Let's just compare a professional olfactory scientist, who thinks about olfaction just going about their day, to a search and rescue dog." Horowitz laughed before continuing, more seriously, "I'm interested in behavior. What are people doing with their noses? What are dogs doing with their noses? We humans are not using orthonasal olfaction predominantly. Dogs simply are. If you look at the high-performing dogs who are trained . . . not to be better smellers, but just to be motivated to do the specific smelling task that you want them to, and to tell you when they have found the thing that you have instructed them . . . they're way off the charts."

She cautioned us not to draw too-definite conclusions from Sobel's study. "That Berkeley study had a couple of interesting features. One is the track they laid continued to exist in the ground. In other words, they laid down a string infused with chocolate scent—a scent already detectable by the typical nose—in the grass. And they kept the string—the source of the odor—there. So the source is still present in the grass, as opposed to any tracking dog who's searching for the odor that's *come off* the source. There's still obviously something in the air—molecules being effused—otherwise, the dogs wouldn't be detecting anything. But for them, the source of the molecules is gone. That's the whole point. So a more useful correspondence would be to lay the string down and then *remove it*—because then you'd be asking them to detect the trace left by the source, not the source itself."

Horowitz also noted on the temporal difference in the responses of dogs and humans. "A number of subjects made it down the route. But the time they take is incredibly long. It's many minutes, I think up to 14 minutes to do the path—if they pass at all, if they didn't just abandon the whole thing." That aside, she remarked, "I think the interesting thing is, [participants] get better. It showed that we have the power to use our noses. And we can get better if we practice, which every perfumer could also tell you. . . . Even so, the dog's apparatus seems miles beyond ours. Their natural behavior is different than ours. I think that is what forms a clear distinction between dogs and people."

Whether we are as good, just as good, or slightly less good in comparison with dogs is not the issue. The point is that the behavioral capacity to detect and trace smells is markedly accurate also in humans. The difference is that humans rarely perform tasks that involve scent tracking; they rarely crawl on the ground on all fours. But that is where most smells originate. Most odors stick close to the surface of things.

Essential to sniff-enabled spatial behavior in olfaction is movement. It is an enactive exploration of distance relations between odor sources and the perceiver.[17] In contrast with how it processes other distal senses, the untrained brain does not calculate the spatial dimensions of an odor source from afar (as it does the temporal delay of an auditory signal, which depends on a stimulus with regular spatiotemporal behavior). In olfaction, the brain receives information about the distance of a source by having the nose in close contact with the stimulus and the perceiver move along its concentration gradient, meaning the distributed chemical concentration of an odor plume.

The computation of this interaction involves a temporal dimension. Finger added: "You have to follow the trail to get to the object. Whereas with vision and taste, you immediately know where the object is, instantaneously. To me, olfaction requires much more temporal integration to get a spatial cue out of it." Here, Aina Puce replied, olfaction "shares similarities with audition in some ways, where a stimulus must unfold in time to be recognized."

Perceptual theorizing highlighting such sensorimotor activity gained traction in so-called ecological theories of perception, enactivism, and embodiment. These theories were spearheaded by the psychologist James J. Gibson in the 1960s and 1970s.[18] He was followed by the biologists Francisco Varela and Humberto Maturana in the 1990s.[19] Today, these theories come in a variety of flavors in philosophical debate.[20] Intellectual differences aside, proponents of these theories warn against the implicit separation of perception, the body, and environment in the analysis of perception. How an organism's body is built and what kind of actions it affords fundamentally structures the content of perception. Accordingly, sensory input and motor output must be viewed as coupled and analyzed alongside each other (although there is considerable disagreement about the precise definition of such coupling). The body of the perceiver evolved to realize perceptions for exploration-oriented interplays of an organism with its environment. It is the chief task of the senses to operate as reliable guides for organismal decision-making. Biological factors consequently have a significant differential impact on how external information is first encountered and subsequently realized as perceptual content.

Behavior fundamentally determines how information reaches the brain and gets processed there. The brain, of course, is part of the body; one of its central functions is to coordinate and integrate bodily signals. Sniffing presents a glaring example of how behavior affects central processing because sniffing behavior orchestrates oscillatory rhythms of brain activity. Neural oscillations are caused by membrane potential changes in neuronal signaling, which can be measured with electroencephalography (EEG). These oscillations represent the activity of neural populations, and their fluctuations mirror alternating states between higher and lower excitatory phases. These state changes underpin the selection of input information. Higher excitatory phases favor the detection of a particular kind of input; lower excitatory phases suppress the detection of the same kind of input.[21]

Sniffing affects how oscillation rhythms in connected brain areas interlock.[22] Don Wilson explained: "We know that there's an entrained oscillation in the olfactory system whenever you breathe." Entrainment refers to the temporal coordination of neural activity, such as the synchronization activity of two neural populations via breathing. "It is respiration-entrained oscillation." He mentioned a recent study by Gottfried and postdoctoral scientist Christina Zelano.[23]

As you're breathing, the activity in all these regions is going along with your respiration in synchrony. It's not the dominant signal, but there's enough that you can pick that out. We know the olfactory system is strongly tied to hippocampus and amygdala and these other areas that are important for different kinds of memory. Respiratory entrainment helps; so when I'm breathing the actual act of inhaling and sniffing is linking those different brain regions together, and I then have this smell. I just turned on my hippocampus simultaneously and my amygdala simultaneously and they're all building this large contextual representation of smell.

Further studies strengthened insight into the dynamic connection between sniffing and response state changes via cell recordings, including the positive link between active sniffing and olfactory learning.[24] It remains inconclusive, however, how sniffing determines intensity.[25] In effect, oscillatory entrainment constitutes a key mechanism by which motor action in the nose determines information processing in the brain.

How can oscillations affect perceptual content? Oscillatory rhythms constitute a form of neural sampling behavior in which input signals are sampled via a process of "perceptual cycling." The cognitive psychologist Ulric Neisser coined this term in 1976. Neisser described perception as a cyclical process in the brain, suggesting that search patterns in foraging behavior filter input information.[26] Alternating oscillation phases mirror the periodic sampling of sensory input and govern the responsiveness of particular brain regions, including their connectivity. Several neural populations are actively competing at any given time. So the brain is primed by its own mechanisms of input selectivity.

The brain is not a homogeneous structure but coordinates several modular clusters. In a sense, it is in continuous competition with itself. Neural populations participate in an ongoing tug of war to reach a sufficient threshold of attention. Attention, however, is a limited resource. Perceptual cycles are a mechanism of active selection to avoid a corollary overflow of signaling in sensory processing. Different behavioral factors, such as sniffing, change the rhythm of oscillations in perceptual cycles.

This coupling of perception with motor action radically underpins the formation of perceptual content: what gets processed and even what reaches consciousness. The underlying idea of active sensing is not modern; the physician Hermann von Helmholtz had formulated a similar understanding in his ideomotor theory in the nineteenth century.[27] Modern are the means to explore this coupling with its precise effects on percept formation.

Why do smell images not have discrete spatial properties like visual objects if olfaction likewise facilitates spatial behavior? With this question, we touch upon the core that ought to define theorizing about perception if it is to be more than a theory of vision. The dimensions and features of perceptual objects, including their boundaries or discreteness, are computational features. These features are a product of neural topology—that is, they are a product of how the information transmitted by neural signals is integrated and formatted. Olfaction, however, does not compute odors in a topographic manner the way vision or audition do.

Neural Space

Both the neural map and what it represents remain comparatively unknown in olfaction. As sensory entities, odors entail multiple dimensions: they communicate certain qualities, come in varying intensities, and are often experienced as pleasant or unpleasant. These perceptual dimensions—quality, intensity, and hedonicity—are measured separately in most sensory performance studies.[28] Moreover, they are considered to map separately onto neural structures.[29] The critical point in olfaction is that the spatial topology of the olfactory stimulus does not map onto the spatial topology of its neural representation (see Chapters 7–8). It is essential to model olfaction in terms of its own computational structure instead of concepts derived from the visual system that may not apply. After all, vision and olfaction do show significant computational differences in feature coding.[30]

The coding of spatiality is vital in this context. Spatial qualities of visual objects (edges, shape, orientation, and motion) are not a direct result of enactive processing—such as eye saccades or sniffing—but a consequence of the computational structure of the visual system. To be sure, organismal movement aids and contributes to the processing of spatial features in all distal senses. Sometimes you may need to move a little to apprehend and assess an object in its distance to you, and equally its shape or size. But such movement is not the matrix that generates and implements spatial features as qualities of perceptual objects. We saw that spatial coding in vision is a result of the topographic projection of signals from retinal cells sensitive to light contrast (center-surround cells) to cells in the primary visual cortex (and their hierarchical integration from simple cells to more complex cells). This setup is responsible for the possibility of viewpoint-invariant features.

Shape coding in vision builds on viewpoint-invariant features in edge detection. So-called T-junctions define the boundaries of objects, and Y-junctions determine areas where surfaces join.[31] The visual system uses such junctions to calculate regularities in perspective. An excellent example is the reconstruction of parallel structures; indeed, optical illusions routinely build on this computational feature. (The Ames room is an example. It is a visuo-spatial illusion of a room that, viewed through a peephole, looks like a cube, whereas its true form is a six-sided convex polyhedron. One corner of the room is farther from the observer but not perceived as such. Consequently, a person standing in the distant corner appears to be dwarfish compared to an equally sized but gigantic-looking person closer to the observer.) The computational structure that underpins spatiality further promotes the idea that visual object recognition operates by stable components, involving the template matching of distinct geometrical shapes (for example, in "Geon theory" by Irving Biederman).[32] Recent findings in the anterior inferotemporal cortex, with neurons coding for a variety of abstract shapes, support this idea.[33]

Olfaction does not show comparable computation regarding the spatial features of its stimulus. Finger explained: "Vision is a topologically oriented sense. You're mapping a three-dimensional world onto a two-dimensional surface and then calculating out a three-dimensional world from that. Olfaction is totally different. Olfaction doesn't give you very good spatial cues, although there are some. The computations involved in olfactory experience are very different than what you would get from vision. So I think vision and olfaction are about as far apart as you can get in terms of central processing." He added, "There may be commonalities in neural mechanisms, but I think how the brain has to treat the information is very different."

There is a big difference between "spatial objects" and "spatial representation." The critical point about neural representation is that it is not primarily about the spatial relations of the distal stimulus to the environment, but a topology created by the sensory system. Because spatiality is immediately associated with vision, these two notions are conflated easily. But they are not the same. This crucial point is best exemplified with the auditory system.

Audition is also a spatial sense. From a phenomenological standpoint, we experience sounds as coming from the left or right side and a particular distance. (With loud noises, you do not need to move your head to figure out from where they are coming. A bang and you instinctively turn your head toward the sound, without thinking.) That is not the kind of spatiality involved in the topographic representation of the stimulus in the auditory cortex. Sounds are carried by the frequencies of audible pressure waves (in humans approximately between 20 hertz and 20 kilohertz). Their topological representation in neural space is a consequence of how the auditory system handles the stimulus, meaning how it breaks stimulus features apart to reorganize them into a discrete neural signal. Its topography in neural space is not a representation of how the auditory stimulus is part of physical space.

The auditory system is a delicate masterpiece. Pressure waves reach the eardrum (tympanic membrane) through the ear canal. Three consecutive tiny bones (hammer, anvil, and stirrup) propagate the vibrations from the eardrum to a little membrane that is part of the cochlea: a snail-like structure in the inner ear from which the auditory nerve transmits the sensory signal to the auditory cortex. The cochlea is an intricate formation. It contains a variety of components, including the basilar membrane that receives sound vibrations via neighboring hair cells tuned to different frequencies. This is where the topography of sound is created.

Imagine the basilar membrane analogous to a piano. Different membrane segments resonate with the sound frequencies hitting your ear (Figure 5.1). No matter how complex the sounds are, regardless of whether you listen to Tchaikovsky or Aerosmith, your basilar membrane lines up these frequencies as detected by hair cells of different length. The resulting linear representation of sound on the basilar membrane is projected via the auditory nerve and maintained in the primary auditory cortex. So if we were to look into your brain and see activations in a specific area of the primary auditory cortex, we would know whether you are experiencing a tone with a high or low frequency. This spatial segmentation is not a representation of the physical stimulus as a spatial object in the environment, but a representation of the signaling components arranged in neural space.

Spatial localization of sound sources in the perceiver's environment is achieved via multiple means, not directly via the topographic coding of the auditory spectrum. The auditory system uses

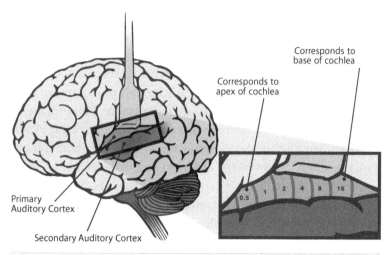

Figure 5-1 Auditory cortex frequency mapping. Illustration of the spatial organization in the basilar membrane. Incoming sounds are separated by frequency, ranging from low to high. This topographic arrangement is projected to and maintained in the primary auditory cortex. Source: Lars Chittka and Axel Brockmann/Wikimedia Commons CC BY-SA 2.5.

multiple cues: next to spectral differences (represented by neural topography explained above), spatial localization is based on the computation of time differences with which the stimulus reaches each ear, as well as stimulus intensity. So the auditory system diverges from vision in how it evolved to code and compute the spatial dimensions of the stimulus. The last example to illustrate this significant difference between physical and neural spatiality is pain perception. Pain perception is not a distal sense. Pain has no representational relation to distal objects and their location in space. The pain you feel when you hit your kneecap is not a representation of the edge of the desk that caused it. Meanwhile, pain is perceived discretely in the phenomenological space of your bodily experience (the pain is *in* the elbow) via localized activities in neural space (designated areas of your brain fire while others stay silent). In an embodied and neural sense, pain is spatial. "The case is analogous in certain respects to that of one pain in your arm being closer to your elbow than a second pain," the philosopher Chris Peacocke at Columbia University noted. "The experienced locations of each pain do really stand in spatial relations, and it can really be true that one of them is closer to your elbow than the other." This level of spatiality, however, does not correlate to objects and their spatial properties in the environment. "Acknowledging this point is entirely consistent with regarding pain itself as a nonrepresentational state."[34]

Spatiality in perception (embodied as well as representations of distal space) is a computational feature. The link between properties of perceptual images and their neural representation requires a model that determines which features are coded, and how these features contribute toward an overall perceptual image. Computational space is governed by stimulus behavior in relation to the observer. Its construction hinges on the predictability of the stimulus, and how the sensory system evolved to facilitate action relevant to the organism.

"There's nothing inherently spatial about the stimulus," Stuart Firestein noted. The epithelium breaks the spatial features of the stimulus apart, "so why ought there be anything spatial about its [neural] organization? It seems to be an oversimplification based largely on, as these things tend to be, work on the visual or auditory system." Linda Buck joined Firestein's assessment of the scattered stimulus: "Smells are not like that. They're kind of distributed."

Spatiality in smell perception is embodied. Intensity coding in olfaction enables assessment of behavior in space; but it is not coding for construction of space, like in vision. Instead, smell partakes in orientation behavior by banking on overarching, nonolfactory computations of external space. Finger concluded: "So there is some spatial information. It's just not highly mapped. I think of it more like having an internal representation of your three-dimensional world. You're just mapping some information onto your internal representation of space."

Therefore, if we want to understand how perceptual content relates to neural representation in olfaction, we must understand what kinds of activities are involved in all of this. Odors are the activity the sensory system performs. That brings us to the first gateway of the system: receptor coding.

CHAPTER 6

Molecules to Perception

Outside your body awaits a multiverse of airborne chemicals. They spark a vast range of different odor qualities and behavioral meanings and are strikingly diverse in molecular structure. After registration by the nose, it is up to the brain to do something with all that information. But what is happening here? Unlike colors and sounds, the connection between stimulus structures with odor qualities is far from evident. We have seen that the stimulus did not readily explain odor. So how does your nose know that the molecule of cis-3-hexen-ol smells of freshly cut green grass, or that the chemical group of esters smells fruity? How does your brain decide whether it got its perceptual interpretation of these chemicals right?

The answer to these questions depends on what we think our sensory systems do when they scan a stimulus for information. The notion of perception we have come to rely on is the idea that our brain conducts an efficient extrapolation process, whereby our sensory systems get to the observable nature of things. Perception as extrapolation conveys the assumption that our senses operate by filtering information from contingent and variable scenarios to detect stable patterns of the world encoded by its physical features. Neural representation, in this context, is an act in which the brain re-presents previously encountered, learned patterns to categorize the currently presented information. In this scenario, sensory perception functions as an information funnel, denoting a process capable of selecting features in a vast environment of distractions and successfully extracting the significant bits. But what are the significant bits? And how does the sensory system represent them?

Here, trouble starts for olfaction. It is not entirely clear. That is not to say that we lack the data. To the contrary, we know an awful lot about the minute details of the stimulus by now. Keen chemists can blow your mind with knowledge about the structural features of odorants. Big companies like Firmenich and Givaudan amass large databases with detailed molecular descriptions to aid in their search for new synthetic fragrances and flavors. An ångström in difference here, an added carbon atom there, and what about this hydroxyl group that donates its electrons to a benzene ring? Knowing such details is so meaningful that it is impossible to get access to these well-guarded and proprietary research databases.

The Missing Link

We do not understand what the olfactory system does in detail with all those features, how the brain makes sense of them as scents. That is astonishing because over the past three decades, we have discovered a lot about the biological foundations of olfaction (Chapter 2). Of course, you sometimes encounter comments about how little we know of olfaction (perhaps the odd sentence in this book has fallen victim to a similar sentiment). But if you think about it, we know a fair amount about the sense of smell today. It is just that we also have come to realize how little we understand of what we know. It is not that the principal components are unfamiliar. To date there may be more structural data than we can make sense of. We can study the molecular specifics of the stimulus in all its chemical glory. We know about the structure of the pathway where this information is processed, including the vast number of receptors and the signal projections to the bulb and cortex. All the bits and pieces are in place. Only the principles of the olfactory cascade remain in debate. What is missing?

Missing is the connective principle, the topology that undergirds the process of perception and integrates its information at different levels. The stimulus, in one form or another, is the source from which our sensory system extracts information. Unclear is how the olfactory stimulus communicates its message. The general pointer to stimulus topology (it's the chemicals!) distracts from really thinking this answer through. Odor chemistry is mindbogglingly complex, and the physical stimulus of smell still has no comprehensive classification. This, we now know, is not a result of the seemingly subjective nature of odor, but the molecular complexity of the olfactory stimulus.

The neuroscientist Charlie Greer at Yale reminded us of the root of the problem: "One of the most difficult things is that we don't understand *the chemistry of the system*. We still don't understand what a ligand is and how it interacts. That is in stark contrast to the physiology of the somatosensory system, where we understand hot receptors and cold receptors and pressure receptors at an extremely detailed level. Or the visual system. Or the auditory system. In many ways, these systems are, at least in my view, comparatively simple compared to the olfactory system."

Talk about stimulus input is ambiguous. Even in vision we are already talking about two different things. One is the distal object, a thing perceived in the distance (like a line projected on the screen). Another is the causal stimulus, meaning the photons hitting your retina. That they are different kinds of objects is evident.

Photons do not have lines or edges. They do not have shape or length. They do not have any of the properties we routinely assign to visual objects. Instead, they act as surface reflections that our visual system uses as a *measure of* distal objects. Our ability to see distal things as spatial connects to the fact that the causal stimulus behaves spatially in its interaction with the visual system (Chapter 5). Visible objects appear spatial to us because spatial dimensions (like distance and size) are determinate of the information that our system extracts from their surface reflections.

So how does the stimulus behave in its interaction with the olfactory system? We won't find an answer by looking at odorants in isolation. We don't even use this approach in models of the visual system, Stuart Firestein emphasized: "We don't worry about the physics of photons for the most part. There is a whole lot of work on photons done by particle physicists. Are they waves? Are they particles? Vision scientists worried very little about that. They were interested in optics, that's about it. But only because they had to set up an optics table to *deliver* the stimulus."

The reason chemistry dominates olfaction is a matter of historical convenience. The twentieth century was the time when chemistry was the best option to study odor, experimentally. This paradigm somehow survived.

"We have this common trope in the field, from molecules to perception," Firestein observed. For decades, the assumption was that there are rules that link chemical input to mental output. Today, olfactory information is still analyzed as encoded in stimulus structure, while the rest of the story, including the receptors, amounts to filling in the molecular details of the biological apparatus. By tracing how the receptors project their input to the brain, we arrive at a more or less linear model of the wiring of the system (like edge detection in vision). This model, however, is valid only if the receptors respond to odor chemistry as chemists model it. That is not the case.

Receptor biology is governed by its own rules. "The trouble with the idea to connect molecules with perception is that it goes from chemistry to psychophysics," Firestein noted. "What's been left out all these years? Biology!" Twenty-five years after the discovery of the olfactory receptors, and a century of stimulus chemistry, the question we ought to be asking is: How does the system work? "Now we have to put biology back in," Firestein argued before pointing out: "But when we put the biology back in . . . *It doesn't fit*. It doesn't fit that nice story of chemical structure to psychophysical perception. There's lots of other stuff going on instead."

The previous chapter revealed that odor processing is not about the distal stimulus as an external object, but a topology as created by the sensory system. This chapter examines why there is a big difference between the chemistry of the stimulus and the topology of its neural representation. Attention now turns to how biology reads chemistry.

The Common Trope

Chemistry presented a plausible starting point for initial scientific interest in olfactory biology. "This is the way olfaction has gone for a long time," Firestein observed. "Because it is called the chemical sense, right? All the molecules we smell are, for the most part, organic compounds. And you know," he shrugged, "there is a whole field called organic chemistry. Naturally, you expect them to take the lead on this. They name these things, they have extracted them, synthesized them. They run that chemical show. It's perfectly reasonable to rely on organic chemists to organize and classify the chemicals that they spend all their time working on. Which we [neuroscientists] don't because we just use them."

Neuroscience did not need to start from scratch. Odor chemistry was already in place when biologists entered the field. "You don't have to believe that this is the final answer," Gordon

Shepherd replied, "but it is definitely a tool for a much deeper understanding. It's almost a list of how to represent the input. To me, the simplest idea—since this is the idea of how the study of most senses occurs—is that you need to know where you are in the field of the sensory input in order then to stimulate the different parts of it. Just like the visual field. And then to know where to go in your system in the brain."

The sheer number of receptors complicated that idea. Richard Axel noted: "If you have a thousand different cells, and an odor activates one hundred receptors, the number of possible combinations is greater than the number of atoms in the universe! So that's a big number, a very big number. This immediately gave you the power you needed to recognize as many molecules as you would ever wish to recognize in your entire existence." That revelation inevitably altered ideas about odor coding.

"It appeared that biology was now possible to do," Firestein remarked. "The idea initially was to try and get the biology and the receptors to fit into what we already thought was going on based on the chemistry and the psychophysics. And the biology should just fit neatly in there. It doesn't work out that way as it turns out. But it's reasonable to think that way or to start that way."

Still, the stimulus remains at the center of olfactory theories. Can modern olfaction, with access to receptor biology, continue to build on structure-odor rules? Comparing past with present insights reveals a hidden shift in ontology.

Over the past couple of years, several articles tried solving structure-odor rules (SORs) with big data.[1] These studies advanced computational models of the olfactory stimulus, utilizing artificial intelligence to mine for clear correlations between chemistry and psychophysics. This approach also marks the arrival of a new generation in the olfactory community.

Testing new tools on old problems, Andreas Keller found, was a no-brainer: "There are these things that are just obvious that should be tried." His collaborator Pablo Meyer agreed: "There's just a couple of obvious things to do. I mean, why not do it?" Joel Mainland thought that tools such as machine learning fueled a generational shift that also mirrored an epistemic break with tradition: from explanation to prediction. Machine learning constituted "a new set of techniques that the field has not absorbed yet."

Computational perspectives promised to crack the code in the nose with more sophisticated techniques, more data, and better data processing tools. Rick Gerkin, from a neuroinformatics view, said: "You can answer a little question here and there, but to answer questions like 'What is the dimensionality of olfactory perceptual space?' and 'How many odors are there?' you need to have large data sets, and large data sets take a long time to collect, they take a lot of money, and most labs doing olfaction and olfactory psychophysics are smaller labs that can't answer those questions."

One central problem with these new computational studies were the data. Leslie Vosshall remarked, "most of the theoretical work [in olfaction] has been based on [this] single thirty-year-old data set. Why has no one done an update?" This old data set is the *Atlas of Odor Character Profiles* (Chapter 3). Andrew Dravnieks, Vosshall continued, compiled "a great list in the early '80s, for use in the northeast of the United States, for people who are baby boomers. But so many of the words on that list have no frame of reference for the people who come to our studies." She added, "Any of these lists . . . they are perishable, highly culturally biased lists, that will work for some specific period in history, for a specific target audience."

Another problem with Dravnieks's *Atlas* is that its psychophysics was insufficient in methodology. Dravnieks had picked those descriptions himself. Computational studies mapping the semantics of "odor quality space" via Dravnieks's verbal descriptors lacked practical experiments involving human psychophysics. In a sense, they had mapped the odor quality space of Dravnieks.

Computational SORs faced the same problem as old SORs: they black-boxed the biology of the system. What if they had real psychophysics data?

A 2017 publication in *Science* by Andreas Keller, Leslie Vosshall, and Pablo Meyer provided just that.[2] This study is notable for several reasons. First, it used concrete, new psychophysical data on human odor responses, taken from an extensive study published prior in 2016 (also by Keller and Vosshall).[3] Second, this data set was massive. The value of human data collection in olfaction cannot be overemphasized. Keller and Vosshall tested forty-nine test subjects, who sniffed and assessed the quality of no fewer than 476 molecules (using nineteen semantic descriptors as well as ratings of odor intensity and pleasantness). Keller and Vosshall tested a wide range of odorants on an unusually large number of participants (for the underfunded field of olfaction, that is). "And it's incredibly boring work," Keller laughed. "You give people a molecule and ask how it smells. You can't do anything less exciting. It's like descriptive science in its purest form. But it is needed. So we bit the apple, and we tested it."

Third, the article is notable because it represented a modern take on scientific collaboration as crowdsourcing. The 2017 *Science* paper put this 2016 psychophysical data set into use with machine-learning algorithms mining for SORs. The study setup was as follows: It started with a public call for participants as part of the DREAM Challenges (an online open crowdsourcing platform for researchers to pose a scientific challenge for others to participate in). The challenge was straightforward enough: find an algorithm accounting for two data sets, one a list of the chemical features and the other the results of the 2016 psychophysics study. An additional, smaller set with chemical data was given to the participants afterward, allowing them to test and adjust their algorithms before submitting the final version for evaluation. Keller laughed: "So this is the challenge: I collected a data set, and we split it in two and gave half to the people. And we were like: This is how these odors smell, predict for us how these other odors smell." Results of the two winning algorithms were published, but the algorithms themselves were not. The winners were Yuanfang Guan, a computational bioinformatician who had won several challenges regardless of the topic utilizing algorithm fitting; and Rick Gerkin, of whom we just heard. It is worth highlighting that Keller et al.'s 2017 article has been the most successful approach to big data in olfaction thus far; it constitutes a benchmark for similar proposals in the future.

Still, an algorithm is not an explanation. The article "Predicting Human Olfactory Perception from Chemical Features of Odor Molecules" provided a strong case of data mining and confirmation of a number of existing hypotheses on relevant structural features. But, at 0.3, its correlation was not sufficiently high. The DREAM Challenges project did not crack the code in the nose.

Its publication attracted the interest of science writers, like Ed Yong, partly for its big data appeal.[4] The study also evoked cautious critique from olfactory experts such as, for instance, Avery Gilbert. Gilbert's concerns did not target this particular study but applied to computational approaches to olfaction more generally. He identified the absence of psychological theory. Verbal descriptors constitute an arbitrary measure to account for the mechanisms of perceptual categorization. Gilbert's review exposed how disunited the field still is, with computational neuroscientists modeling the sense of smell in a manner markedly separate from that of cognitive psychology.

Gilbert emphasized that the olfactory space remains unknown: "So if one wants to predict what molecules might smell of sandalwood or citrus, one would have to retest all 476 molecules on another forty-nine sensory panelists using the new list of descriptors, then rerun the computer models on the new data set."[5] Why even these nineteen descriptors? Vosshall replied: "The reason our paper used nineteen was just that we didn't find most of the other 127 descriptors applied to the molecules we were using. I'm sure you could use others." The remaining problem, in Gilbert's view, "is that words that are useful in an olfactory lexicon occur at different levels of cognitive categorization." In response, Keller and Meyer emphasized not to view their 2017 work beyond its objectives. It aimed to provide and demonstrate the application of computational tools for odorant design, not for a systems theoretical account of olfactory processing. And so they did just that.

Structure-odor rules, as the go-to strategy in modeling olfaction, are not theory-free tools. Trouble arises from the viewpoint of wet-lab neuroscience. Biology is not data to derive from an algorithm. Biological organization is the *explanandum* (the thing to be explained), whereas algorithms may aid in the derivation of the *explanans* (explanations).

Firestein thus considered these new tools heuristics, not explanations: "There's potentially valuable information in there. I think these machine-learning studies are good leads." So, he cautioned: "They are published like final results, but they're not final results. They're loaded with artifacts. There are all sorts of false positives." It would be imprudent to rule out structure-odor breakthroughs via machine learning, but it has not worked as yet. Why this is the case matters profoundly.

Prevalent in recent computational models is the treatment of biology as a proxy, as a stand-in connecting the chemistry of the stimulus with the perception of the human subject. Mainland argued that this is feasible: "If you want to go study one receptor really carefully and figure out how that receptor responds to an odor, that's great. But it's really a huge pain to do. Instead, use these methods, like [the DREAM Challenges], where you take a molecule and learn what features correspond to perception. In theory, if you have enough data, you'll learn exactly what Stuart is learning. You're using a different set of features, but you could eventually infer everything that he can infer." Mainland paused. "Would you eventually want to know what the receptors are doing? Absolutely. Is it possible that we can figure this out without ever looking at receptors? Yes, it's possible. We don't need to know what the receptors are doing to figure out how to map structure to percept. The current models are basically doing that. And they work relatively well. They're noisy, but they work. You don't have to know every single step along the way to make the jump. It can be a black box."

Keller agreed: "I think about it as a triangle thing: the molecules and stimuli, then you have the pattern of activated receptors, and then you have the percept. You could predict from the physiochemical features what receptors it activates, and then you could predict from what receptors are activated what the perceived odor percept is. You just cut out that middleman and move over to black box off the receptors."

Gerkin went a step further: "We already know these receptors. We know about how many receptors there are. We broadly know how some of them are tuned, and we know something about how they interact in the bulb. But my point is that you can throw all that in the garbage. You can develop a theory of olfactory perception without knowing any of that. My hypothesis is that you can use psychophysics and make measurements to make strong predictions about the grand perceptual space, what the shape of the space is, and how stimuli mix in that space." This optimism may be premature. But is it misguided?

Black-boxing the receptors is bound to fail. Even the most powerful tool cannot avoid the problem of theory-ladenness, the consequences borne from the selection of premises and evaluative criteria. Consider an alternative example. Imagine using strictly morphological criteria for inferences about the mechanism of heredity. The resulting model would be based on correlation, not causation. SORs, structure-odor rules, whether gained by classical chemistry or big data, similarly circumvents the biology of the system, the causal grounds of feature selection, and integration by the olfactory system. Modeling SORs, with whatever technique, offers a *lead* to a hypothesis but not the actual mechanism. SORs do not equal the principles of stimulus processing and perception.

It is vital to make this difference clear. Stimulus chemistry is often framed as coextensive to odor coding. Yong's astute article about the DREAM Challenges project, "Scientists Stink at Reverse-Engineering Smells," is a good example. If you read carefully, you find one notion missing: receptors. What is omitted in most popular accounts that introduce the challenge of modeling olfaction are the receptors interacting with the chemical stimuli, the receptors determining what features get selected. But these receptors are the key to understanding how the olfactory system turns molecular features into neural patterns of information. Back to our alternative example of heredity: what gets taken for a solution here is a morphological description without the mechanism of transmission that determines the units of transmission.

Mainland raised the critical issue: "The only case where it matters [to include biology] is when you get something out of the biology that's not in the things that we're using." Do we have sufficient reason to think that knowledge of biology would lead to a different model of the stimulus in odor coding?

Indeed, we have.

The Black Box of Biology

It all starts with the receptors. Their importance for theories of odor coding cannot be exaggerated. Chapter 2 detailed that olfactory receptors are G-protein coupled receptors (GPCRs), situated on the cilia of the olfactory sensory neurons in the nasal epithelium. Cell distribution in the epithelium is random (although rough gene expression zones in the epithelium exist).[6] These cells continuously change as receptors die and renew. The olfactory system has a constant turnover of sensory cells. (The epithelium is the only part of your body exposing nerve cells to the outer world: a fantastic target for infections. If your epithelium did not renew itself routinely, you would be unable to smell anything after two or three colds.)

For Greer, that is what distinguishes olfaction: "This is the only central nervous system, mind you, where populations of sensory neurons die on a regular basis and are replaced by new populations of sensory neurons—who then correctly send out their axon to the right part of the olfactory bulb to converge with other similar axons." The fact that the system rewires regularly shapes how it interacts with an irregular, unpredictable stimulus. The interface with which the nose scans for odors is under constant construction. And that is not the only notable feature.

Olfactory receptors, as the interface of the olfactory system, actively structure stimulus input—so much so that subsequent theorizing about the neural representation of odors must begin with knowledge of the receptors and their binding behavior, similar to input models in vision or audition (Chapter 5). While all sensory cells are selective, however, olfactory receptors stand out for a couple of reasons.

First, there is stimulus-receptor affordance, the things the system can do with the properties of the physical stimulus. Color vision deals with a low-dimensional stimulus: electromagnetic wavelength. Color receptors, cones, are dedicated to specific chunks of the visible light spectrum. These receptors operate in an additive and subtractive fashion in combination with each other. This results in a straightforward stimulus-quality model; say:

$$a = n$$

"Red light" has a wavelength spectrum from about 390 to 700 nanometers.

Such a model further allows for well-defined feature calculations of color combinations:

$$x - y = z$$

"White light" minus "green light" results in "pink."

While odor receptors are tuned to specific features and act in a combinatorial fashion, this is where similarities end. The physical characteristics of odorants are considerably different from visual input and do not afford the same kind of calculations. The olfactory stimulus is multidimensional in its molecular makeup. Stimulus-receptor space in olfaction is not defined by accumulative combination, as in vision or audition.

Aldehydes, specifically chain aldehydes, present an excellent example to illustrate this difference. Chain aldehydes come in different lengths of carbon chains. (These organic compounds are popular materials in perfumery; indeed, Chanel No. 5 was the first perfume that consisted almost exclusively of synthetics, namely a string of different aldehydes.) Aldehydes of different lengths have different smells. The C8 aldehyde is perceived as fatty, the C10 aldehyde as citrusy, and aldehydes with longer chains come off as floral. Unlike for colors and wavelengths, however, no accumulative model links the number of carbon atoms to odor quality. Besides, it is impossible to apply chemical explanations of aldehydes to other chain odorants—say, alcohols with different carbon chain lengths (ranging from four-carbon butanol's clinical smell to six-carbon hexanol's green note, to eight-carbon octanol's aromatic odor).

Odor coding does not afford a predictive stimulus-response model in the manner of "for any odorant with a carbon chain the model holds that a chain of C8 + another C = results in a cherry scent." That is just not how it works. The essential difference between the low-dimensional stimulus in vision, or audition, and the high-dimensional stimulus in olfaction is that an additive scale does not capture the coding of the latter.

Greer contrasted receptor coding in olfaction with the auditory system: "I guess you could argue that, because the basilar membrane is a continuum in response to high-frequency versus low-frequency tones, there is an opportunity for a combinatorial code there as well. As you play a chord of music, you're going to be stimulating different parts of it, and that will lead to the perception of the music. But I don't think it has the open endpoints that we have in the olfactory system." In the coding of the olfactory stimulus, there is no transience in the range of one key feature. "There's a continuum of tones that you can see putting on a map," Firestein added. "In olfaction, you don't see that kind of continuum. There's no continuum between aldehydes and ketones. Or any other kind of chemical group or classification."

Any model that aims to map perceptual odor space onto stimulus space must begin with the fact that odor coding is not linear or accumulative. Odor receptors deal with several thousand different molecular parameters in no particular order of continuity or scale. Therefore, there is no uniform way to carve the physical space of odorants "at its joints" like visible wavelengths or audible frequencies. Olfactory receptors make sense of about five thousand molecular parameters, including stereochemical configuration, molecular weight, hydrophobicity, functional groups, polarity, basicity, and so forth. This is what is meant by high-dimensional stimulus space.

Odor receptors determine the range of chemical features translated into a signal. But they do not split up the stimulus into uniform, regular chunks (as do the visual cones). You will not end up with a receptor group for one chemical property—say, carbon chains—and another group for polar surface areas. Instead, receptors pick out different features. Moreover, they vary in their range of features (next to feature combinations). Say you have a receptor responding to the polar surface area of a ring structure—but only structures of a certain size, not the polar surface areas of ring structures in general. Now multiply such combinatorics several hundred times, up to the thousands! Chemical features distinguish stimulus space. But these features are not carved up uniformly across the receptors. And so this mosaic coding allows for some degree of data fitting when it comes to structure-odor rules. It does not support the predictability of SORs.

If the receptive range of cones to wavelengths defines color, why don't we define odor by the receptive behavior of the olfactory receptors? Neurobiologists agree that odor percept formation builds on receptor patterns.[7] What remains unquestioned is whether receptor patterns indeed match traditional odor chemistry. Could the study of receptor behavior overthrow the premise of stimulus-response models?

The answer is yes. Two recent studies by the Firestein lab, in 2016 and 2018, tested a deceptively simple question: Would the receptors classify the stimulus differently than a chemist would?[8] Chemists group odorants according to significant chemical groups and functions. Firestein's team measured receptor responses instead (an approach known as medicinal chemistry in pharmacology). The idea behind the experiments was simple, yet no one had considered it. "It was Zita who came up with this idea," Firestein said, crediting his former postdoctoral researcher Zita Peterlin. "I think, in her mind, she sees these chemical structures like no one else. A bit like in the movie *A Beautiful Mind*. She sees patterns in these molecules that others do not."

Erwan Poivet, who continued Peterlin's project after she left for Firmenich, summarized the idea: "Organic chemistry will rely on stuff like: What's the functional group of your molecule? What is its size? What is its length? How many double bonds, is it polar, or is it a-polar? All these different features. And that will be the way chemists classify the molecule. But maybe this is not relevant at all for biological systems, such as the olfactory system. Maybe your receptors don't care about there being an acid or an ester. Let's say you have an aldehyde and alcohol. Chemically they're pretty different. But for olfaction, you still have oxygen in both of them with a double bond. Maybe that's what the receptor is tuned to. Maybe it doesn't care if you have a hydrogen or another carbon." The Firestein lab threw a range of molecules at a mouse epithelium. They recorded and analyzed the receptor responses to check whether the stimulus preferences in receptor behavior matched the stimulus classification in organic chemistry. The answer was no. The olfactory receptors did not match stimulus chemistry as a trained chemist would. What that means is that the receptors follow their own rules. Without the details of receptor biology, SORs and big data are left with force fitting.

Figure 6.1 shows the differences in how chemists and receptors group the chemical similarity of odorants. At the top of the figure, you see how odorants 3 and 5 are most similar according to the principles of organic chemistry, subsequently followed by odorants 6, 2, 1, and, last, odorant 4. At the bottom, you see how odorants 5 and 6 are most similar according to the receptors. Meanwhile, odorants 1 and 2 form a similarity group, independently of odorants 5 and 6; and odorants 3 and 4 are closer to the group of odorants 5 and 6 than odorants 1 and 2. "We chose to look at molecules that, if you look at their chemistry, were quite different," Poivet explained. "We looked at cyclic molecules with benzene or heteroaromatic rings to see if we can find something to relate them. And the way they were classified by chemistry was very different from the way we can classify these molecules after we force them into neurons."

The real experiment was more sophisticated than the above suggests. Poivet laughed: "We had olfactory sensory neurons in a petri dish. And we inoculated [introducing a substance into the tissue], one by one, all these odorants. Every time we have a neuron that responds, we can follow its calcium sensor GCaMP [a fluorescent protein highly sensitive to calcium activity in cell activation]. We found that there were cells that respond only to one

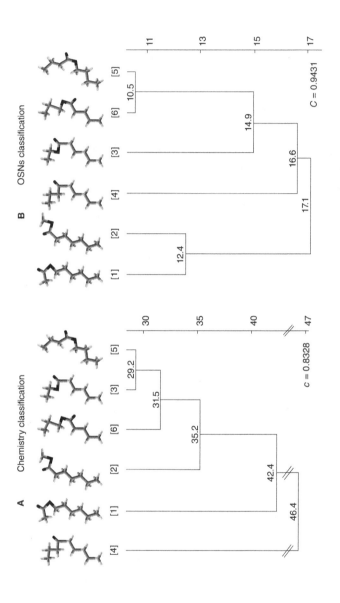

Figure 6-1 Hierarchical clustering analysis of tested esters. Comparison of odorant similarity. The left tree shows how analytic chemistry grades the chemical similarity between ketones. The right tree shows us how the olfactory receptors determine chemical similarity. These classifications differ significantly in their classification of chemical similarity (explanations in text). Source: Reformatted from Poivet et al., "Functional Odor Classification through a Medical Chemistry Approach," *Science Advances* 4, no. 2 (2018), fig. 3 CC BY-NC.

molecule, cells that respond to two molecules, cells that respond to every molecule. It was a pretty good mix of everything! We looked for some patterns. Although they were all ketones [in the 2016 study; the 2018 study added esters], they differed by the number of carbons in their cycle, also polarity and, more interestingly, the polar surface area of the ring." (The polar surface area is the sum of the surfaces of polar atoms, such as nitrogen, oxygens, and hydrogens.)

It turned out that the olfactory receptors could not care less what analytic chemistry dictated. "The way [ketones] would be classified by organic chemistry," Poivet argued, "would be the size of the ring as a first way to separate them, and then the composition of the ring as a second way. Do we have a nitrogen, or an oxygen, or a sulfur atom? And that would be your subfamily in the big family of a five- or six-carbon cycle group." The receptors had other preferences. "In olfaction, the classification we have was very, very different. The size does not matter at all. Neither does the cycle composition. What mattered was actually the polar surface area. This is where you have the electric charge in the three-dimensionality of your cycle, which actually accounts for the fact that the neuron will accept the odorant as a ligand or not."

In less technical terms, Poivet and Firestein's study had two notable findings. First, they found that the priority and hierarchy of features by which chemists and receptors determine chemical similarity differed. Some features that classical chemistry highlighted were of little interest to the receptors. Chemists and receptors had a different idea of what defined odorants as structurally similar. Second, the receptors responded to chemical features that hadn't even been predicted, or were even on the radar, of previous SORs and big data studies.

Poivet nodded: "The pattern we found was that if you have an olfactory receptor that accepts the lowest polar surface area of the molecules, and also a molecule with a larger polar surface area, your receptor will accept, at least for these cyclic molecules, all the molecules with a polar surface area in between. No matter their size, like if these are bigger in their cycle or not. That was pretty interesting because—just with organic chemistry—you couldn't have predicted this!"

How biology makes sense of chemical similarity, therefore, diverges from the ideals of chemists. This changes how we should arrive at a theory of smell perception. Just like in cryptography, you have to have the right key to break the code. Everything else is word salad, regardless of whether it yields a few sentences that might make sense. To understand what a neural signal represents hinges on knowledge about what stimulus features the signal encodes. Consider an analogy. When physicists define the term "gravity" it matters whether they interpret it according to the framework of Newton or after Einstein. Both theories describe gravity as a field. But Newton saw gravity as a force on top of absolute time and space (as separate notions), whereas Einstein defined gravity as a curvature of spacetime. When you now model chemical similarity in olfaction, think of this as a similar paradigm shift.

Consequently, two principles of receptor-stimulus interactions ought to center in olfaction theory: first, the combinatorics afforded by the multidimensional stimulus, and second, chemical similarity according to receptor behavior. These two features highlight why odor biology is not a black box linking stimulus chemistry to perception. But there is another noticeable characteristic of olfactory receptors, one that ultimately shapes the neural representation of odor.

The Blind Homunculus

The brain represents what it is shown by the receptors. It does not "see" the configuration of external odorants but deals only with signals from the epithelium. Informational units, the signaling

pieces that perform a coding function in perceptual object formation, thus are determined by the mechanisms and patterns of receptor interaction, not the chemotopy of the distal stimulus.

Two fundamental mechanisms shape the signal that reaches the
brain: *combinatorial coding* and *inhibition*. Combinatorial coding
splits the information of the physical stimulus into several independent signals. Inhibition means that some parts of a stimulus
can block the activity of another (such that the receptor pattern
of a blended mixture is not a simple addition of activation signals
by its components). These mechanisms, taken together, render the
notion of chemotopy (as the neural representation of external stimulus topology) untenable.

The consequences of combinatorial coding for signal transmission and neural representation is twofold. First, the signal is underdetermined because it is crosscutting and overlapping. Several
odorants interact with a receptor (and vice versa).[9] Moreover, different molecular features can activate a receptor. What the activity
of a receptor represents, therefore, is not indicative of a specific
property or microstructure. Second, the signal is further ambiguous because the binding preferences of receptors are uneven. Not
only do code receptors exist for multiple bits of chemistry, but they
also have different tuning ranges in their combinatorics. Each receptor type responds to a particular range of features. Some of
these receptors are broadly tuned, interacting with a vast number
of various features and odorants. Others are highly specific, responding to a smaller number of parameters. You have to know
the behavior of a receptor for an idea about the kind of information and scope it signals.

At the receptor level, the external signal gets thoroughly scrambled. Say receptor type R1 detects a specific functional group of
odorants, while another receptor type, R2, only recognizes chain
structures of a definite length (that is, four to six carbon atoms).
At this point, the informational content of the olfactory stimulus

is split into numerous bits and pieces across the receptor sheet. All this activity is mixed together on a single spatial plane.

Combinatorial coding has significant implications for the coding of mixtures. It entails that different odorants, in combination and under more natural conditions, could overlap in the receptors they activate. That's important when you look at mixture perception. Firestein explained: "You put the mixture on [expose the tissue to the stimulus], and you see a whole bunch of cells light up. Then you put each odor on separately, and you look at what cells light up. Of course, if you add up the numbers individually to the number that you see by the mixture, it's less." Firestein thus warned about the limitations of monomolecular stimuli. "Everything we usually do is monomolecular. Dissociate the cells [separate them from vessels and cell aggregations], put this odor on, and see what you'll find lighting up. Put another odor on, you see some other stuff. But that's very unnatural because everything we smell in the world is a blend of a few things up to hundreds of things."

A general theory of odor coding should be built on the principles of mixture perception. For one thing, stimulus information at the receptor level is no longer linked to individual odorants (as discrete external objects). Cell activation in the epithelium appears as a spatially distributed pattern. Activation patterns are randomly distributed and overlapping. What we end up with is a field of feature combinations, where the information of one stimulus (odorant O1) is not topologically discrete from the activity induced by other odorants (say, O2 and O3) encountered in parallel with O1. Distal stimulus topology ends up scrambled on a single plane, such that the interpretation of olfactory signals is contingent upon the mechanisms of the sensory system, not the external configuration of the stimulus.

Consequently, the brain cannot identify single odorants in a mixture by their combined receptor activation patterns. Consider

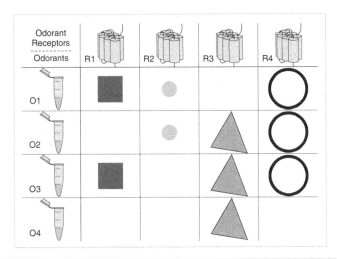

Figure 6-2 Hypothetical example of combinatorial odor coding at the receptor sheet. A mixture containing both odorants O1 and O2 overlaps in its receptor (R) coding with a mixture that consists of odorants O1 and O3. Likewise, a mixture containing both odorants O2 and O3 overlaps with a mixture involving odorants O1 and O4. Source: © Ann-Sophie Barwich.

the following hypothetical example. Imagine a receptor activation pattern R1-R2-R3-R4. In principle, this pattern could be caused by different sets of molecules as a consequence of combinatorial overlap. Figure 6.2 shows what this might look like. Effectively, in mixture recognition, it is not possible for the receptors to unambiguously determine the individual components.

This poses an intriguing challenge. How does the brain really know what kinds of things it encounters? Think about it by comparing the underlying issue to the story of *Flatland*.[10] Flatland is a fictional world, a two-dimensional place with two-dimensional beings. One day, Flatland's inhabitants see a three-dimensional object moving through its two-dimensional plane. Such an encounter with a three-dimensional object appears as a two-dimensional pattern on the plane. Now think about this plane as the receptor sheet and your brain as a Flatlander observing the receptor patterns. Suppose a spherical object like a ball moving through Flatland. It would start as a small point that grows into a circle, increasing in size before it turns back into a small point until it vanishes. Now assume the pattern of another object moving through, a spinning top. It starts out as the same! Just by looking at the plane, it is impossible to tell whether the two-dimensional patterns of Flatland are a representation of a sphere or a spinning top. A similar case holds for olfaction, in which receptor patterns, forming an odor object, can be caused by a set of different odorants. Different odorants can generate the same mixture pattern.

Hold on a minute, you may say. Sure, receptor combinatorics implies that some mixtures have overlapping activation patterns. Still, in principle, odorants involved in mixture patterns could be determined by excluding double activations on the receptor sheet. It is a bit more complicated, perhaps, but possible. Some ambiguity in these patterns would remain, but we could derive a general theory of odor coding from individual odorants rather than having to deal with mixtures. Notwithstanding the challenge for the brain, making sense of scrambled receptor data does not stop here, because odorants in a mix might also block each other.

This type of stimulus inhibition at the receptor level is unheard of in the other senses. It might be unique to olfaction. It is not known to happen in vision or audition. It is not known in taste or touch or any other sensory modality, as far as we can tell. "I don't know of any other sensory system that does this." Firestein sounded excited. "Green photons activate green cones, but they don't inhibit blue cones or red cones. There's color opponency and all that, but that's upstairs, right?" He pointed to his head. At the receptor level, "there'd be no mechanism for that. You can get a red photon at a high enough luminosity to activate a green cone a little bit. But they don't block; there's no inhibition."

Could there be inhibitory processing right at the first step of odor coding? A recent study by the Firestein lab indicated just that.[11] As the majority of olfactory research shifted its focus toward the big questions in central processing, the Firestein lab continued to probe the receptors. Firestein did not think that the relevant functions were sufficiently understood. "So we got interested in blends," Firestein said. But how could researchers determine if odorants blocked each other, instead of merely overlapping in their receptor activation (as a result of combinatorial coding)? Whatever mixture you throw at the epithelium, "it's less than you would get out of a simple mixture." Firestein explained. "Because there are some receptors that are seeing both or all three of those features, so you're double counting."

The answer arrived in the form of a spectacular new microscope, SCAPE.[12] SCAPE stands for swept, confocally-aligned planar excitation. Firestein laughed, "anything for a catchy acronym. Essentially, it's based on a light sheet kind of microscopy. But it's a rapid scanning light sheet so that you can record many cells in a volume of tissue and very rapidly—quite an improvement really."

SCAPE opened a new window of experimental opportunities, amassing terabytes of data. It made it possible to scan an entire living, moving fly—for example, researchers can puff some odors at it while looking at its brain *in action*. Tissue samples larger than flies and larvae, like sections of the brains of mice, can also be scanned. The novelty of SCAPE was that it allowed scanning an entire intact tissue section while also recording single cell activity, both at an incredibly high speed and with high resolution.

"We took a hemisected preparation," Firestein detailed. "So we have the [mouse] head in a dish, perfusion in and out, and we can image a large swath of the olfactory epithelium at a depth. We can get down to a volume, down to a depth of about 180 or so microns. But we can also do it at the single-cell level. So you can also, when you want it, get single-cell resolution. It's like a combination of doing single cells and EOG [electrooculography]." With SCAPE, one can determine which cells react specifically to what odor to distinguish the patterns. And all that can be done in intact and active tissue, not in dissociated cells or fixed brains cut into slices. Firestein noted: "The obvious thing to do with this would be blends or mixtures to see the code." The data, collected by Firestein's graduate student Lu Xu, are beautiful. You can now look at a tissue section and see how the entire preparation responds to a stimulus.

The findings yielded two major surprises. The first surprise was that odorants acted as both agonists and antagonists. "Apparently, in the mixture, one of the components is acting not only as an agonist but also as an antagonist at one or another receptor," Firestein said. This means that an odorant O1 can modulate receptor activity such that cells, activated by other odorants, say O2 or O3, show reduced activity or no activity at all when presented with a mixture containing both O1 and O2 or O1 and O3. Moreover, this odorant O1 does not act as an antagonist per se, but only acts in combination with specific other odorants (which may act as antagonists to different odorants as well). Antagonism, therefore, depends on the particular combination of an odorant with other odorants in a mix and is not a feature of the odorant per se. Firestein confirmed: "We've done a couple of mixtures now, and we've never found an odor that acts only as an agonist or only as an antagonist."

Inhibitory effects in olfactory mixture perception have been known as perceptual phenomena in psychophysical tests.[13] These effects had not been linked to a mechanism, however. Was inhibition an effect arising at the periphery and/or by central processing? There had been some earlier reports on inhibitory effects at the receptor level in olfaction.[14] What surprised everyone now was the sheer amount of inhibition. Inhibition was not a phenomenon affecting only the odd one or two receptors. "It's pretty widespread!" Firestein emphasized. "We find that in a mixture of three odors—when you look at the three odors, and then you look at the mixture—we can see as much as 20 to 25 percent inhibition. It's a lot of inhibition that goes on. If you look, for example, at a cell that you see is dominantly activated by citral, and then you look at those cells in the mixture, you find that as many as 20 percent of them are being inhibited."

There was a second surprise, with even more enormous implications. Results further indicated *enhancement* effects, next to inhibition in mixture coding. Enhancement means that some cells, which showed little to no response to any of the individual odorants, suddenly responded actively to a mixture of these odorants. Firestein knew this was important. At first, he admitted: "I can't quite make sense of that part yet." The study continued while this book was in progress. Shortly before manuscript submission, Firestein emailed that they had linked the effect to an explanation: allosteric interaction. This mechanism—roughly!—states that a ligand (like an odorant) binds to a specific site at the receptor (the allosteric site) and thereby alters that receptor's activity. In other words: odorants *modulate* how an effector binds other odorants. As an example, a receptor R1 does not bind a given odorant O1 administered individually. If this odorant O1 is presented in a mix with odorant O2, however, then odorant O2 attaches to the allosteric site of the receptor to modulate its activity such that it now binds odorant O1. Lu Xu and Firestein tried different variations of mixtures, with mixtures that contained components in equal as well as unequal concentration. The enhancement effect remained robust.

Allosteric interaction had been well known in pharmacology, yet it had never been observed in GPCRs. Xu et al. had an answer to this puzzle: "That it has gone undiscovered in other Class A GPCRs [one of the six GPCR classes, grouped by sequence homology and functional similarity] is perhaps not surprising since they comprise a much smaller family of receptors than the olfactory receptors and there is much less variation between them."[15] The size and genetic diversity of olfactory GPCRs, and their range of structurally diverse ligands, indeed makes them an excellent model to study other GPCRs, a subject of great relevance in pharmacology and drug design.

But what is the function of such inhibitory and enhancement effects specifically in odor coding? Xu et al. suggested that it serves discrimination and identification of complex blends. Consider the effects of combinatorial coding in olfaction: "Making conservative estimates that any given odor molecule can activate three-five receptors at a medium level of concentration, then a blend of just ten odors could occupy as many as fifty receptors, more than 10 percent of the family of human receptors. This will result in fewer differences between two blends of ten similar compounds." So you'd end up with indiscriminate odor activity in comparisons of more complex mixtures (which often contain dozens, even hundreds, of odorants). Patterns of odor activity become less and less distinct; this is also because receptors have overlapping sensitivities. How can the brain differentiate between different complex mixtures given these enormous levels of receptor activation and pattern overlap? You need to reduce receptor activity to refine discrimination of different mixture combinations. Inhibition and enhancement mechanisms serve this purpose.

Ultimately, this finding manifests a paradigm shift for a theory of odor coding. It shows that the receptor code in mixtures is fundamentally different from receptor codes of monomolecular stimuli. The idea of a linear, additive combinatorial model of odor coding, like that in vision or audition, breaks down completely. You cannot crack the olfactory code without understanding receptor behavior.

Odor coding modeled on the combinatorial scheme is, to a sufficient degree, underdetermined. Different odorant blends may end up with the same receptor representation, meaning spatial distribution of odor signals cannot disambiguate olfactory identity in mixtures. Compare the idea with a familiar notion in the philosophy of science: the "underdetermination of scientific theory by evidence" (proposed by the French physicist Pierre Duhem, extended by the American philosopher Willard Van Orman Quine).[16] This notion states that different, even incompatible, theories can accommodate the same sets of observations. The same observational data, thus, can be read wholly differently depending on the interpretational framework. For example, the fact that the sun rises in the east to set down in the west is compatible with both geocentric and heliocentric models of the universe: same data, different models. We have now seen that a similar principle characterizes the combinatorial coding of odors at the receptor sheet. So how does the brain know what really happens outside the nose? How can the nose accurately tell which odorants it encounters? And what could possibly be the function of such indeterminate coding?

Without a model starting from receptor behavior, we cannot understand how the brain makes sense of smell—what it signals and represents via its neural activity patterns. Xu et al. thus noted that their findings about receptor coding carry further implications for central processing. They observed that the brain recognizes smells via pattern recognition, not combinatorial coding and topographic mapping: "Together with the recent work in piriform cortex suggesting a lack of topographical representation, there is abundant motivation to consider alternative coding strategies that also account for the presence of receptor modulation at the first step of olfactory discrimination." The next two chapters will unravel the details of this claim to propose such an alternative.

For now, let's conclude that mixtures yield effects that are not predictable from models that determine odor coding via individual components. While the precise mechanisms underpinning these effects remain part of the ongoing inquiry, we saw that a general theory of olfaction must start from receptor responses to the stimulus, not chemical topology defined by traditional chemistry. Steven Munger replied: "What the brain is eventually going to see may be completely unrelated to what that individual component would have done."

Where Molecular Science Meets Perfumery

The olfactory system evolved to evaluate odors in context, not in isolation. That is the first crucial step to understanding the mechanisms of odor coding, which continue in central processing. Molecular clouds are not discrete, separable objects since odorants also mix with their environmental background. The nose thus measures odors in relation to each other and as part of an olfactory landscape. That implies two tasks: the assessment of complex mixtures with each other (same or different), and the evaluation of components as part of a complex mix (including salience, and figure-ground segregation). The fact that the nose can detect individual volatiles with remarkable precision in this context does not mean that that is its central computational principle.

Mixture perception is where molecular science pairs with knowledge in perfumery. Olfactory receptors show significant suppression and enhancement effects in mixture coding. These molecular effects, while surprising to scientists, have been a long and well-known perceptual phenomenon among perfumers.

Consider the toilet revolution. (Yes, you read that right.) The Bill and Melinda Gates Foundation recently joined forces with Firmenich, the biggest fragrance producer in the world, to find a solution to the stench of public toilets in rural areas with low or no water resources.[17] Water-free toilets are a problem in sanitation: what neutralizes a lot of the stink is water. Without water, public toilets turn into chambers of olfactory torture, containing an unbearably condensed combination of feces, urine, body odor, food, and smoke. Forget waterboarding, really. Naturally, people prefer to defecate on fields in the fresh air, resulting in a disease threat and potential source of communal infections. To facilitate behavioral change, Firmenich and the Gates Foundation worked together to make the odor of these toilets more appealing.

Next to social impact, this work has fundamental research implications. It finally presents a link to connect perceptual effects to a molecular basis on which to model olfactory coding. Matt Rogers, involved in this project, said: "This project is a development of malodor counteractants, which are receptor antagonists—molecules that block the receptors from the malodors that were identified and put in the latrines in Africa. We delivered this list of antagonists to the perfumer, who was supposed to build a fragrance with this antagonist molecule."

What perfumers know, and molecular science has started to explore, is that many perceptual effects in olfaction link to the blending of odor (Chapter 3). Some odorants act as antagonists that suppress the perception of other odorants in mixtures. What turns an odorant into an antagonist, however, often depends on its combination with other odorants in a mixture. The sensory system does not "sum up" its stimulus; it often relies on principles that come to light only in mixture coding.

This intersection of molecular with perceptual expertise is also an opportunity for psychology to reenter discussion. Psychological theorizing can contribute in the formulation of computational principles that correlate odor coding at the molecular level with observable perceptual effects (Chapter 9). Marion Frank, at the University of Connecticut, argued: "The field should look at the olfactory system as it operates in more natural situations. Namely, what it is doing with as many as three-four distinct chemicals, at once, each of which changes in intensity over time." Frank's number is not arbitrary, but links to the Laing limit (named after a series of studies by David Laing in the late 1980s).[18] Laing discovered a limit to how many individual notes a person, trained and untrained, could identify at once in a complex mixture, a perceptual "cap." The Laing limit kicks in usually at three individual notes for untrained noses, and about three to five in the case of expert noses, indicating a general limit to sensory processing, not an absence of training. This gives us the first fundamental clue to odor coding. It builds on pattern recognition, and this pattern recognition is not determined by the coding of individual odorants but by how the system handles them in combination.

The nose *samples*—and the brain *measures* mixtures. This idea of measurement comes into play in two ways already at the periphery.

First, there is the calibration of the system. For the brain to act as an environmental measure it needs a background against which to evaluate change, detect novelty, and recognize saliency. Remarkably, your olfactory system does all that without being distracted by the present odorous background. That's because your nose habituates and adapts quickly to odorants, although not at equal speed. This uneven adaptation of odor receptors exacerbates the scientific study of mixtures. At the same time, uneven adaptation is a determinative mechanism in mixture perception.

Some mixture components are suppressed as a result of selective adaptation after a period of time so that nonadapted elements appear more prominent.[19] Consequently, the same mixture is perceived differently the longer people smell it. Plus, adaptation rates between people differ. Thomas Hettinger argued that selective adaptation explains how our system is tuned to perceiving odors as part of mixtures. "We take, say, three components of the mixture, and then we add a fourth component. So we take the three components, we sniff [this mixture] a few times; we 'adapt out' some of that background. Then we immediately sniff the mixture with four components. The fourth component is perceived above the background of the other three components. You can show that you can extract out information about individual components of a mixture." He emphasized: "A combination of mixture suppression and selective adaptation allows you to recognize the components in mixtures." Mixture coding is where chemistry meets psychology via biology. Frank agreed: "Combined studies of the well-known psychophysical phenomena of 'mixture suppression' and 'selective adaptation' bring experimental control over the natural workings of the olfactory system."

Second, there is the computational scaling of olfactory information, starting at the receptors. Such scaling involves a measurement of "how much" and "in what proportion." To evaluate the disposition of chemical information in context, the olfactory system breaks apart the sampled information into multiple pieces before reconstructing an odor image. That image, we know now, is not the sum of its molecular parts. How does the brain compute the odor image of a mixture from its multiple, different individual components? Again, the clue lies in mixture coding.

The computation of odor images concerns the ratio in which the system detects odorants in mixtures. Recent studies suggest that the olfactory system weighs the ratio of odorants as a form of pattern detection. Hettinger and Frank analyzed concentration measures, using the concept of the odor activity value (OAV).[20] In this, they worked parallel with the chemist Vicente Ferreira at the University of Zaragoza.[21] Frank explained: "This concept is defined as the ratio of odorant concentration to its threshold value. With a modest number of assumptions, it was concluded that the ratio of the identification probabilities (P1 / P2) is approximately equal to the ratio of the odor activity values (OAV1 / OAV2). This transformation is important because it helps to establish the contribution of components in flavor and fragrance mixtures that are often described by odor activity values."

Does odorant ratio determine odor images? Terry Acree provided further experimental proof. His lab recreated the aroma of "potato chips" using only three key odorants.[22] Acree's synthesis of a complex aroma from a handful of key odorants does not lead to a deflationary explanation of odor, reducing odor quality to a few physical parameters. None of the key odorants on their own smelled of potato chips: methanethiol smells of rotten cabbage, methional smells of potato, and 2-ethyl-3,5-dimethylpyrazine smells of toast. The crucial discovery was that the configural image of "potato chip" did not depend simply on the list of ingredients but was linked to the ratio in which the three key odorants were put together.

Calibration and scaling are integral to measurement. They are also central to olfactory coding, linking perceptual effects to a molecular cause. The primacy of proportion and ratio in mixture composition is another phenomenon known from perfumery (Chapter 9) as well as biology—Steven Munger noted: "Complex mixtures of chemicals are very precise not just in their chemical composition but with the ratios of those components. The olfactory system needs to pull them apart so that it can recognize the individual components—but do it in such a way that key aspects of the mixture are retained in the pattern of output to the brain. The pattern is encoded by the nervous system in such a way that the animal can make an appropriate behavioral response."

The upshot is that it may not merely be the "what" but the "in what relation" that underpins the coding and computation of odor quality. What neural mechanisms allow the olfactory brain to receive input and operate in this manner, to sample and measure then represent and map the variable composition of its chemical environment?

Topology in Neural Representation

Receptor coding showed that the brain does not model the olfactory stimulus like an analytical chemist would model a molecule. We need to go beyond stimulus chemotopy to understand the neural representation of odor. Studies on the mechanisms of mixture coding in this chapter paint a complex picture. But the brain has to have some idea of what reaches the nose. Receptor patterns are not the sole or final answer. By some measure, the brain arranges this vast mosaic of receptor activity. Beyond the receptors, however, neural activity encoding mixtures does not furnish us with transparent stimulus-response maps. The olfactory stimulus cannot be captured on an additive scale since its coding and computation is not additive, not in the bulb (Chapter 7) and even less so in the olfactory cortex (Chapter 8). From the brain's point of view, the same receptor activations (the brain's observations) can be generated by multiple distal objects (the physical stimulus).

How the brain interprets receptor patterns turns into an intriguing puzzle. The question no longer is how the brain knows that, say, cis-3-hexen-ol smells of freshly cut green grass. Instead, it is how the brain assigns meaning to overlapping, nondiscrete receptor activity in response to odorants. So how does the brain organize the scrambled receptor activity into neural assemblies and perceptual images? By which principles does the brain make sense of mosaic data from the receptor sheet? These questions lead us to the olfactory brain in the next two chapters.

Pandora's box has been opened.

CHAPTER 7

Fingerprinting the Bulb

Complex biological systems, like the senses, are rarely understood by a simple correspondence between the properties of their elementary components. That said, convergent arrangements like receptor types in the olfactory bulb, with its discrete spatial activity, are markedly conspicuous. How much of this structural arrangement is determinative of function? "You never know," Stuart Firestein cautioned. "In neuroscience, maybe in a lot of biology, you look at the way something exists, and you go: 'I can see that this would serve this function very well.' But often you don't know whether it came to look that way because it serves that function very well, or because it's a solution to a developmental problem. It has nothing to do with the way the system functions as a mature system. It's just that's the easiest way to wire it up. I think that is the case here [with the olfactory bulb]." What does that imply for our understanding of function in the olfactory brain?

The olfactory pathway appears straightforward. It provides a nearly uninterrupted route from the air to the cortical core. Just two synapses mediate the neural representation of chemical information in the environment. That is the most direct pathway of any sensory system. (In vision, two synapses will not even get you beyond the retinal layer!) And yet we still don't fully understand how the olfactory brain represents sensory information. Its simplicity is deceptive.

How is this disorganized mosaic of information at the epithelium translated into a robust percept in the course of merely one to two synapses? This question is key to contemporary developments in neuroscience because the olfactory brain does not reveal itself as an abbreviated version of other sensory systems. Functional localization, the central paradigm of sensory neuroscience, is associated with stereotypic (that is, genetically determined and reproducible) stimulus mappings in sensory cortices. It turns out that this model may not apply to olfaction. The divergence of neural organization in olfaction from other sensory systems is the topic of this, and the following, chapter. The workings of the olfactory bulb are indicative of the hidden complexity in the computation of olfactory signals.

One, Two Synapses Straight into the Cortex

At first glance, the olfactory system presents a shallow route of three basic stages. Information is first picked up by the receptors situated on the cilia of the sensory neurons in the nasal epithelium. The activation patterns across the receptor sheet are spatially irregular, as Chapter 6 illustrated. Receptor signals are then sent to the olfactory bulb at the inferior frontal lobe of the brain, where they are collected in so-called glomeruli (spherical neural structures). In the bulb, we suddenly find spatially discrete activation patterns. This is possible because of a peculiar genetic feature of the system. Each glomerulus collects the signal from all neurons expressing one specific receptor gene. (Technically, as far as we know from mice, neurons expressing one receptor gene project to about two glomeruli—albeit sometimes more and sometimes fewer.) Glomeruli are innervated by mitral cells. This is the first synaptic interface.

The mitral cells, named after their similarity to a bishop's hat, pick up the signal from the receptor neurons and fast-forward it to several areas in olfactory cortex. Next to the cortical amygdala, entorhinal cortex, and olfactory tubercle, the majority of axons of these mitral cells project to the so-called piriform cortex, the largest domain of the olfactory cortex, which connects to several neighboring cortical regions also beyond the olfactory system. Here, olfactory signals quickly intermingle with areas involved in multiple other processes, such as cross-modal interactions (olfactory tubercle), decision-making processes (orbitofrontal cortex), memory (hippocampus), and affect (amygdala). This is the second synaptic interface.

Ramón y Cajal had highlighted this characteristic of the olfactory pathway already at the turn of the twentieth century. His drawing of the olfactory system offers early insights (Figure 7.1).[1] Cajal reasoned that the shallow pathway of olfaction served as an excellent model for general studies of the brain.

Cajal's recommendation was forgotten in twentieth-century brain research. The reason was partly methodological; the olfactory stimulus was tricky to administer and hard to control (Chapter 1). It also seemed impossible to determine a receptive field of the nose, similar to Kuffler's work on the retina in vision (Chapter 2). In hindsight, this makes sense given the wide variety of receptors and in light of the modern challenges of odor coding. Still, after the receptor discovery, why hasn't the neural code been cracked over the past three decades? Cajal's observation makes it sound straightforward enough.

"That's because Cajal did not understand the molecular complexity that was there," Charlie Greer replied. The field still faces a lot of unanswered questions when it comes to the details of the system, noted his colleague Dong-Jing Zou, in Firestein's lab: "For

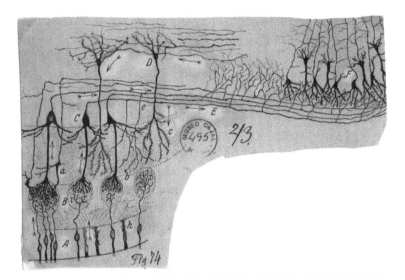

Figure 7-1 Cajal's drawing of the olfactory pathway. It shows how odor information is mediated by two synapses: the first across the glomerular layer (the spherical structures on the left side) and, second, when the signal is projected to the cortex (on the right side). Source: Courtesy of the Cajal Institute, Cajal Legacy, Spanish National Research Council (CSIC), Madrid, Spain.

example, how many mitral cells project from the olfactory bulb? Are they all the same? Only one major cell, or multiple cell types? Nobody has carefully studied that." Randy Reed agreed: "It's a tough question to answer. We know they're not all identical. But we don't know why they're not identical."

Most puzzling about the olfactory brain is that it builds an exquisite topography of odor signals in the bulb, only to be abandoned right after the first synaptic interface. "You have this beautiful map," Richard Axel remarked. "It's one of the most beautiful maps of the brain. It's conceptually beautiful, as well as aesthetically very pleasing." But the cortex immediately forgets about this map. Olfactory signals in the piriform cortex end up thoroughly scrambled, their spatial distribution is largely arbitrary

(Chapter 8). "So that beautiful tightly organized structure in the bulb all of a sudden is in many instances discarded."

The olfactory bulb presents a challenge, one that is not recognized immediately: its spatial arrangement is far from self-evident—for two reasons. The first reason is that this apparent odor mapping in the bulb does not play any further role in the processing of olfactory signals and their computation into odor images. The olfactory cortex differs from other primary sensory cortices in its lack of topographic organization, paradigmatic of neuroscientific investigation of the senses. The second reason concerns the idea of an odor map in the bulb itself. What is being mapped? Besides, how evident is the premise that the spatial patterns in the bulb are, in one way or another, really a map? The topography of the bulb stands on much more brittle ground than previously thought.

Deceptive Simplicity

The olfactory bulb is the neural structure of the olfactory system that's possibly been studied in most detail. And like renewed attention to the retina in vision, recent insights into the bulb challenge preconceived notions about both its structure and its function.

Overhaul of orthodox views starts with its size: the olfactory bulb was long dismissed as a receding structure in higher-order mammals, especially humans. During evolution, and according to the popular psychologist Steven Pinker, olfactory bulbs "have shriveled to a third of the expected primate size (already puny by mammalian standards)."[2] This opinion remains widespread. It does not hold up to scrutiny, however. In a recent *Science* article, the neuroscientist John McGann at Rutgers argued that human olfactory bulbs aren't that small.[3] McGann questioned: "What do you mean by size in the first place? Do you mean proportional comparison, interspecies comparison, or density in terms of the neurons? There are so many ways in which you could look at structure-function relationships!"

Size, like any measure, depends on scale. Proportionally, human olfactory bulbs are smaller in extension. The bulbs, however, are comparable in their number of neurons. One may equally say that human olfactory bulbs did not shrivel up, but that the human brain grew in size. Charles F. Stevens at the Salk Institute looked at the scaling principles of microcircuits to demonstrate a general conservation of neural structures in olfaction across species.[4] Others found that the olfactory bulb presented a notable exception to a generally conserved brain scale, making the bulb's size unpredictable and independent from the rest of the brain.[5] The reason remains unknown.

What this amounts to is the challenge that the move from structure to function poses its own problems. So how shall the function of the bulb be conceived? Analogous to the visual system, must the function of the bulb and its receptive fields be modeled like the retina, the thalamus, or primary sensory cortex? Opinions vary. Historically, the bulb was compared to the retina by Cajal and, later, by Gordon Shepherd. The neuroscientist Leslie Kay, at the University of Chicago, said: "Gordon's argument rests on the dendrodendritic synapse." (We get to that in a minute.) By contrast, and in a friendly dispute, Kay explained, "Murray Sherman and I wrote a paper in 2007, comparing the circuitry of the olfactory bulb to the thalamus."[6] The paper argued "that you have the same connections and properties of neurons." Kay hoped that the thalamus community would pick up this idea "because of the things that we've studied in the olfactory bulb, in its circuitry. We can predict what things will change the system in what ways to a larger degree than the thalamus community can."

Without insight from the information processing mechanism, structure-function mappings remain underdetermined. Such a mechanism may not require or build on topographic arrangements. The spatial activity of the bulb need not be an expression of the computational principles of this mechanism and link to other explanations.

The essential elements of the bulb's anatomy were discovered in the late nineteenth century when Camillo Golgi published a morphological description of the cells in the olfactory bulb of a dog in 1875.[7] Testing his new silver staining method, Golgi's elegant image of the bulb was the first complete stain of an entire brain region. Figure 7.2 shows Golgi's depiction of the bulb in astonishing detail. What immediately strikes the eye are the various layers of the bulb. It is a densely populated area, host to a plethora of different cell types. First, there is the glomerular layer (A in Golgi's image) where all the olfactory nerves from the epithelium terminate. Next, and near the mitral cells, we see the smaller tufted cells that likewise receive olfactory signals via their dendrites innervating the glomeruli, and project to the cortex (layer B).

Before these signals are sent out far away into the cortex, they become subject to local processing within the bulb. So we do not deal with an unmediated mapping of receptor activity in the bulb. This local processing involves lateral inhibition, meaning that excited cells can inhibit (reduce the activity of) their neighbors. That is a useful feature of any system that wants to prevent activity from spreading uncontrollably. It thereby sharpens the neural representation of a signal by delineating the active from the nonactive or less active parts. This computational feature is likewise known to exist in vision and other sensory systems.

At first, things in the bulb resemble what is happening in the retina. Governing the process of lateral inhibition in the bulb is a peculiar type of neuron: granule cells (layer C). Unlike the typical nerve cell, a granule cell does not have an axon and consists only of a cell body (soma) plus dendrites (just like the amacrine cell in the retina; Chapter 2). Its dendrites bridge the distance between

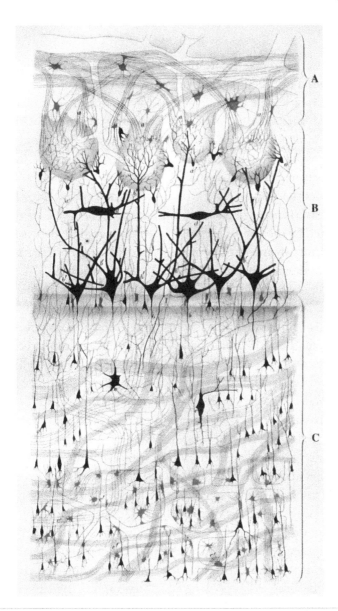

Figure 7-2 Golgi's stain of the olfactory bulb of a dog, depicting its different anatomical layers. We see (A) the glomerular layer with its spherical neural structures; (B) the mitral cell layer of large cells with tufted dendrites, including a depiction of horizontal cells now known as tufted cells; and (C) a thick granule cell layer stretching into the cortical domain. Source: C. Golgi, "Sulla fina struttura del bulbi olfattorii," *Rivista sperimentale di freniatria e medicina legale* 1 (1875): 405–425; reprint (Reggio-Emilia: Printer Stefano Calderini 1985).

mitral cells to coordinate their activity. Granule cells are small yet abundant; about a hundred granule cells latch onto one mitral cell. Granule cells and mitral cells form loops in what Shepherd termed "microcircuits": mitral cells excite granule cells which, in turn, inhibit the mitral cells. This sequential, self-inhibitory activity within the bulb takes place via "dendrodendritic interactions, entirely without axons."[8]

A lot is happening right after the first synaptic interface. It is as if the bulb is managing its own tempo, instead of having its activity dictated by the flow of incoming receptor signals. Now, these microcircuits are interesting because they determine both spatial patterns and temporal sequences of signaling in the bulb.

Zooming into the structure of the bulb, we are confronted with several kinds of cells of different shapes, sizes, and functions. The "shell" of glomeruli, for instance, is composed of juxtaglomerular cells (small interneurons, central nodes communicating between sensory or motor neurons and the central nervous system). Next to the "primary circuit" of mitral and tufted cells, we find a zoo of interneurons in the different cell layers of the bulb, including periglomerular cells (PGs), external tufted cells (ETs), and short axon cells (SAs). These cells form various microcircuits involved in intra- and interglomerular communication. Notably, the study of interneurons in the bulb continues to yield surprises (short axon cells were recently discovered to be possibly the first kind of interneuron connecting to the sensory input layer).[9]

How does knowledge of microcircuits contribute to a functional picture of the bulb and the question of stimulus mapping? These details are crucial in understanding how information from the receptors gets "delayed," "integrated," "synchronized," and "amplified" across the various structures of cell layers in the bulb, according to painstakingly detailed research on bulbar microcircuitry (by researchers including Michael Shipley, Zachary Mainen, Bert Sakmann, Gary Westbrook, Ben Stowbridge, Jeffry Isaacson, Thomas Cleland, Mark Wachowiak, Nathan Urban,

Gordon Shepherd, and Charles Greer, among others).[10] In other words: How much information coming from the receptors is represented by the activity of the bulb? How do coactivated glomeruli interact (plus, is this crosstalk linked to topography)? And is the spatial pattern in the bulb principally determined by stimulus input or by parallel computational processes (like inhibition or top-down), so much so that the same odorant under varying conditions could show different patterns?

Imagine you administered a mixture that consists of two different odorants, A and B. Regardless of suppression effects at the receptors, the next question is whether the signal of mixture AB, broken into several separate signals at the receptor sheet, is processed by linear addition or nonlinear recombination. Look at the hypothetical scenario in Figure 7.3.[11] Let us assume that when individually administered, odorant A activates a specific group of

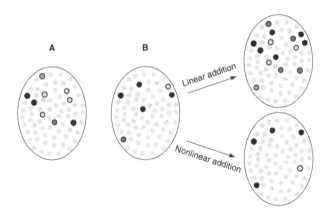

Figure 7-3 Two possible mechanisms of stimulus integration in mixture processing in the olfactory bulb. Mixtures might either be represented by linear addition of glomerulus activity (top right) or by selective combinations of glomeruli (based on inhibition mechanisms) (bottom right). Source: P. Duchamp-Viret et al., "Olfactory Perception and Integration," chap. 3 in *Flavor: From Food to Behaviors, Wellbeing and Health,* ed. P. Etrévant et al., Series in Food Science, Technology and Nutrition (Cambridge, UK: Woodhead, 2016), fig. 3.3 © 2016 Elsevier Ltd. All rights reserved.

glomeruli {G1–G8}. So does odorant B, which overlaps in its activation of one glomerulus G1 {G1; G9–G13}. When mixed, the activation patterns of these odorants together can either be additive {G1–G13} or selective in the activation of glomeruli {G1; G6; G9; G10; G13}. Either option is possible and would depend on factors such as inhibition mechanisms or some signals being too weak to surpass a threshold.

The recorded image of the bulb's activity is not the full story when it comes to the analysis of the constitutive mechanisms involved in its expression.

Breaking Down the Bulb

Remarkable about the bulb is how it organizes the seemingly haphazard, distributed activity at the epithelium into spatially discrete clusters, like a fingerprint or "brainprint" for each smell. Pause for a moment with that thought. After all the combinatorial chaos at the receptor sheet, the bulb suddenly looks like a precise map for each odor, a route directly from receptor activity to neural representation. The scientific literature routinely refers to this arrangement as a stimulus map, in analogy with that of the visual system. Chemical features of the stimulus seem to be mirrored by patterns of discrete spatial neural activity. Different activity patterns should, in principle, account for different stimulus features. Is the bulb a map representational of odorants or odor chemistry? Unfortunately, it's not that simple.

We are looking for an orderly representation of selected physical features from the environment, or some discrete computational expression thereof. An odor map should express its input in one way or other. In relation to the bulb, three interpretations of such a map are available: rhinotopy (systematic representation of receptor position), odotopy (distinct spatial activity correlated with odorants), and chemotopy (spatial representation of chemical features of odorants).[12] Whether any of these options applies is far from evident or even intuitive.

In contrast to the review literature, which reports on the "chemotopic" or "topographic" character of the bulb, understanding among olfactory scientists is less settled. "I was always a little bit unsure of [the idea of an odor map]," Firestein admitted, "because I couldn't imagine, frankly, what that map would look like. What would you map about an odor if you linked it to the bulb, would you put aldehydes next to ketones, or would you put esters in between them? There's no way to answer that because there's no rational reason for doing any of that."

"I don't think that the map of the bulb has to mean anything, really," Linda Buck replied. She considered that the bulb displays a spatial mapping of receptors. "But I think that its relationship to perception or anything else is quite doubtful. I've long thought that the organization of the bulb may arise indirectly by the developmental mechanisms that evolved to get neurons with the same receptors to synapse in the same place. Because it could be beneficial in allowing for low-level signals from thousands of neurons to be integrated into a far smaller number of bulb neurons. It may be important for the sensitivity in the detection of odors at low concentrations, but that map itself doesn't mean anything."

Firestein's former postdoctoral researcher, Matt Rogers at Firmenich, played devil's advocate: "I think there is a chemotopic map, in the bulb anyway. The question is, what is its purpose, back to Stuart's point. Look, I was trained under Stuart, I drank most of his Kool-Aid." Rogers laughed. "I think why we lose faith in the functional relevance of a chemotopic map is that we do not see it yet in the pure form." That said, Rogers clarified: "I say chemotopic map because reading the papers, that's what is seared in my brain as a way of describing it. But I completely agree [with the skepticism]: I don't know if there's a map . . . it's probably not a 'map.' If it's not a map, what is it?"

"I think it's a question of semantics," Charlie Greer jumped in. "There is no question that axons coming from sensory neurons that express the same odor receptor converge into glomeruli are at least broadly distributed in neighborhoods within the olfactory bulb. I will refrain from calling it a topographic map, but there is some *specificity* there."

There are several reasons for abandoning the idea of an odor map in all its three interpretations. The approach of strict chemotopy was refuted in Chapter 6, which showed that receptor mechanisms do not mirror chemical topology. Nonetheless, the principles of receptor binding in the epithelium are different from the mechanisms that determine glomerular organization in the bulb. Does the bulb correlate with the chemical topology of odorants? The answer is also negative. The high-dimensional stimulus cannot be mapped onto a two-dimensional surface of the bulb, not even by a fraction of its features.[13] This finding applies to mammals as well as insects. Studies of the latter have been exceptionally clear on that issue.[14] There is no discrete spatial arrangement of chemical features such as "aldehydes here, ketones there, and esters over there." There is chemical specificity in glomerular activity, but not chemotopy.[15]

What about rhinotopy (in analogy with retinotopy in vision)? While we find rough genetic zones in the bulb, its organization is far from comparable to the topography of the visual system. The special feature of retinotopy in the visual system is that neighboring cells have similar receptive fields. That is indeed the key to its entire representational organization (Chapter 2). At first sight, this seems to be the case for the bulb, too. Each odorant appears to have a "fingerprint" in terms of its activity. Additionally, glomeruli activated by a particular stimulus appear clustered together. At second sight, however, neighboring glomeruli aren't actually more similar in their responses than distant glomeruli—namely, when tested with a variety of odors instead of a few similar classes.[16] John Carlson at Yale University has shown in flies, for instance, just how locally diverse glomerular responses to odors are.[17] Besides, activations of local glomerular clusters to similar odorants aren't "continuous" either but reveal "gaps."[18]

Perhaps, the answer is odotopy. It sounds like the most solid option. If individual odorants appear to have a "fingerprint" in the bulb, and there is at least some overlap in the clustered responses of glomeruli to similar odorants, then spatial activity in the bulb might not be organized by neighborhood, yet it may constitute a map as a coarse, distributed yet discrete, signature pattern. This option is preferred by Shepherd next to Venkatesh Murthy at Harvard.[19] It would yield a functional map. This model comes with a critical condition, however. To count as a functional map, these patterns also must be stereotypic, meaning generalizable across members of a species, if not even across different species.

Stereotypic functionality implies that a pattern can be arbitrary (what input is located where has no intrinsic or systematic value); however, it is repeatable (a particular input is precisely, or roughly similar, *here* in all cases). In other words, functionality is hardwired; it presents an invariant spatial expression of computational principles, including lateral inhibition. So how does the bulb know *where* to put its glomeruli? Ultimately, a stereotypic organization is genetically predetermined.

Organic systems regularly exhibit structures that do not necessarily build on particular functions but merely mirror their development. Could this be true for the olfactory bulb? In one way or another, signaling activity in the glomeruli relates to molecular features of odorants, and the bulb forms something like the receptive field of the olfactory system. This fact notwithstanding, the bulb is primarily a display of receptor activation, not stimulus chemistry. Thomas Hettinger agreed: "It's not a mapping of odor.

It is a mapping of the receptors." This distinction is crucial because it means that the bulb is a consequence of receptor genetics. The following exploration of the historical rise of research on the bulbar map, focusing on its genetic and developmental basis, will cast light on its causal foundation, asking: Is the bulb prewired?

A Fingerprint for Each Odorant

Targeted scientific interest in the olfactory bulb emerged when topographic modeling was at the heart of advances in neuroscience. Gordon Shepherd, leading the way into the bulb from the early 1970s onward, recalled this experience. His motivation was to show that the olfactory system was not so different from other sensory systems: that it was not, as many scientists had believed at the time, the odd one out. Shepherd recalled: "Eventually, we found the creational patterns representing odors! These basic mainstream properties that have been shown in the somatosensory system, in the visual system, and the motor system, and so forth. I think it had to be true also of smell."

Firestein remembered his time as a postdoc in Shepherd's lab: "Gordon was a strong proponent of this idea that the olfactory system was part of mainstream neuroscience. That its mechanisms—however it worked or whatever we were going to find out about it—would fit into what we knew about neuroscience and would fit into the existing models. That was not necessarily a common view at the time. There were a lot of people who felt that olfaction was this idiosyncratic, somewhat quirky system, whose rules were somehow different from the rest of the nervous system, especially the other sensory systems—which is why it was so hard to learn about, to tackle, and to experiment on it. Olfaction was special, in one way or another. Gordon, on the other hand, was quite the opposite. He felt that whatever we learned—whether it was in the periphery, or centrally, or anywhere in the olfactory system—

would make sense in the wider context of the brain and neuroscience in general."

"This was a very difficult point to get across," Shepherd added. "The olfactory people, aided and abetted by the fragrance people, insisted that this was special. I remember one of the IFF [International Fragrances and Flavors] people saying: 'Well, you'll never be able to do any credible physiology on specific kinds of odors because they will always be contaminated.' As organic chemists, they were very aware of how hard it was to prepare even odor-free air to use as a vehicle for you to put different odors in. If an aroma was due to minute amounts of a particular compound—that is always true of a compound—then how could you say that you were working on odor A when it might be B, C, D, or E? This made it very difficult to believe that you could do standard sorts of physiology on the olfactory system compared with the other systems."

These challenges did not prevent Shepherd from trying, and he was proven right. He found some exceptionally distinct and spatially discrete activation patterns in the bulb.[20]

The critical piece remained missing, however. To confirm that the bulb's activity was a map of the stimulus, the olfactory community needed the receptors, the input interface that determined the interaction of the stimulus with the olfactory system. Firestein clarified: "Very little of that [neuroscientific work] was odor-driven. They were interested in delineating the wiring diagram of the olfactory bulb. What they cared about is what was connected to what, and in what way. In what direction were the signals moving? A lot of that work could be done actually without the use of odors." This was not an uncommon strategy in sensory physiology. "A tremendous amount of what we know about the wiring diagram of the retina is done without the use of light," Firestein emphasized.

Once it was clear how the signal was structured at the receptor level, the expectation was that it should be mirrored in its projections throughout the brain. With Buck and Axel's discovery, the necessary piece of the puzzle finally entered the field.

The next steps seemed reasonably upfront: connecting stimulus chemistry to the bulb's patterns by targeting the receptors. A coding fever befell the olfactory community. But receptor coding turned out to be a more significant challenge than expected. Imagine you have several hundred, up to a thousand, different receptor types. Receptors are activated in a combinatorial fashion and randomly distributed in the epithelium. Axel summed up the challenge: "How would the brain know which of the cells have been activated by a given odor?"

The answer arrived promptly in 1996. Peter Mombaerts, then a postdoc at Columbia with Axel and now directing his lab in Frankfurt, found that the olfactory system employed a peculiar trick to deal with random input from the receptors.[21] Each sensory cell only seemed to carry one receptor type—although, according to Mombaerts, doubt can be cast on the generality of the one receptor-one neuron doctrine today.[22] Axel explained the trick: "Turns out that there are a thousand receptors [in mice] and all cells that make the same receptor, despite their distribution in the nose, send the process through the skull into the first relay station of the brain where they all converge on a fixed point"—the glomerulus.

It was an engineer's dream. The one receptor-one neuron doctrine reduced the erratic complexity of the receptor sheet concerning its subsequent wiring. Because the olfactory system has an immensely messy setup, Greer pointed out: "If you try to conceive of the projection from the olfactory epithelium back to the olfactory bulb it is, without question, *the* most chaotic pathway in the central nervous system. You have like 1.2 million or 1.3 million cells on each side of the nose, and then each of the cells has its own original point of origin within the olfactory epithelium. Then its axon has to get back here [to the olfactory bulb], where it converges with other axons expressing the same odor receptors."

Without the axonal convergence of receptor types, it sounded impossible to figure out just how the olfactory system could make any meaningful distinctions between stimulus inputs.

New experimental possibilities opened up in an instant. Researchers now could trace signals from the receptor cells in the epithelium directly to spots in the bulb where they would converge. Axel remembered: "We took a probe for an individual receptor, and we said: 'Maybe we can examine the projection of the sensory neurons in the epithelium in the brain by probing for RNA-encoding receptors in the brain.' We reasoned that maybe some of the RNA gets out into the projections, into the axons. So, we did something called in situ hybridization [where you use a strand of complementary genetic material as a probe to localize specific tissue selections] and simply looked for receptor RNA in the brain. And we indeed saw it in dots. Each receptor identified a different locus in the bulb. Those loci were then anatomically demonstrated to be glomeruli by the Mombaerts experiment. It was a crazy experiment that first time. But it worked!"

The Mombaerts experiment carried a vital implication: there should be a pattern for each odorant. A fixed code of glomeruli would represent each odor. Say, as an example, the odorant citral activates a specific set of glomeruli {G1; G5; G6; G204}, while musk ketone activates another set of glomeruli {G5; G6; G30; G50; G400; G420}. If you were sniffing these odorants, and we looked at fMRI scans of your brain, recording your perception of the two mixtures, we would see a markedly different neural pattern for each.

"That was the beginning of the idea of the glomerulus as a functional unit," Firestein concluded. "By functional unit, the idea was that each glomerulus was dedicated to some set of smells and that it faithfully ordered odor quality by giving it some spatial organization."

Interest in olfaction began to shift from receptor behavior to central processing. "Finally!" Shepherd proclaimed. "Now we're beginning to get a robust number of laboratories working so that they can compare and compete."

Swiftly, the hunt for the neural code behind a chemotopic map took over. The idea of chemotopy was omnipresent in the field. Firestein remembered: "This resulted in an explosion of articles on the olfactory bulb and the spatial arrangement of activation patterns as chemical feature maps. A number of people, based on that discovery, decided that this made sense because now you could imagine some sort of a map, a way where you'd map odor quality—that is the receptors that an odor bound to—onto a central structure. There were several people, in particular Kensaku Mori in Japan, who really jumped on this idea."[23]

Leslie Vosshall replied: "The olfactory glomeruli were a very important concept. Some of the early insect people noticed that there was an organizational principle there, anatomically. Take a moth. There's this huge proportion of the sensory periphery of the male moth devoted to just smelling the female—this whole set of cells in the male antenna that's just about the female. If you follow those cells to the brain, you find this huge expansion of glomeruli that are all about detecting the female. People [were] looking at odor activity in the olfactory bulb, where you start having different patterns of glomeruli lighting up, depending on the density of the odor, and the concentration of the odor, and if it's a blend."

One odorant after another was matched with its particular patterns of bulb activity.[24] The findings showed "different patterns for different odors," in Shepherd's words. "That was an incredibly important series of discoveries made by many labs," Vosshall added. "What seemed important is that you could have this two-dimensional sheet that also has a time that encodes concentration and, in some way that was unclear, odor quality. Some early simplistic works were saying that, in the olfactory bulb, all you need to do is, once you increase the carbon chain length, you have this orderly marching of activity across this array of glomeruli and olfactory bulb. That's probably wrong. Certainly, in insects, there's no clear organizational principle, and I think increasingly [this applies to] the vertebrate olfactory bulb. That's not how it works. It's much more abstract."

Research on the bulb stagnated; the concept of chemotopy remained opaque. Firestein concluded: "As long as we're looking for a spatial map based on chemistry, like aldehydes are here, and ketones are there, or the esters are here, we won't be able to find that [map]."

The hunt for a map obfuscated a deep-seated question: What of explanation? The thing about bulb activity is just that: glomerular patterns are *patterns*. As mere patterns they do not, as Shepherd noted, exhibit an internal logic: "Pattern recognition, that's almost a discipline of its own!" In one way or other, a map of the bulb necessitated a modified, receptor-based model. Receptors, however, follow their own logic.

Stereotypic Representation

If olfactory sensory neurons act like vehicles, carrying the receptor signals to the brain, receptor genes may best be understood as their drivers. So what makes these drivers tick? The chief engineering principle of the bulb is that sensory neurons with the same receptor genes coalesce in glomeruli. This might be a unique feature of olfaction. Greer highlighted: "Each of those cells follows a very independent pathway based upon the point of origin in the epithelium. That's not true of any other sensory system that we know of in the brain." Firestein confirmed: "It's not to my knowledge repeated anywhere else, in any other system where G-protein coupled receptors of this type actually govern the projection where the axons go, at least directly."

Firestein explained: "The discovery that the receptors also had something to do with how the axons [of the sensory neurons] got to the bulb was a most exceptional finding because G-protein coupled receptors had previously never been implicated in axon guidance or targeting. So now you have these receptors doing two things! First of all, there is this remarkable class of receptors. This huge family expressed in the epithelium. Then, in addition to that, depending on what receptor is picked, this governed in some way where a neuron's axon went to the bulb—a finding I find, to this day, curious." The real significance of this finding has yet to be unraveled.

In the meantime, the idea of a map gathered an experimental life of its own. The fundamental hypothesis was that the bulb's placement of its glomeruli is genetically prewired. A fixed genetic map of the glomerular organization was assumed to hold the clue to the organizational architecture of the bulb by underpinning its computational structure. This premise rests on sufficient reason. The topographic representation in other sensory systems, such as vision or audition, is genetically determined. Zou mentioned: "In the visual system you have a pattern. The stimulation is more defined in a topographic arrangement." Postnatal experience further may refine these sensory maps at a later stage. But their principal organization is instructed by genetics, making it stereotypical: "They're topographically invariant from organism to organism," Axel said.

Doubts about the stereotypic wiring of the bulb arrived with further insight into receptor genetics. How did all of these olfactory sensory neurons know where to go? Moreover, how did these cells find their genetic twins with the same receptor? "So that became a question for a while," Firestein said, "as to how these receptors did it." We still do not know. To be sure, this state of not knowing is not because we lack a general theory of axon convergence. Instead, the process of axon convergence in the olfactory system seems to operate by a different mechanism. "Everyone says: well, it's the odor receptor," Greer noted. "Certainly, the odor receptors are correlative. But no one has been able to show that odor receptors are responsible for axon coalescence and how they mediate that process."

How do neurons usually find their targets? The general explanation involves axon pathfinding in neural development. During growth, axons find their destination via a trail of chemical gradients. Here, axon projections are highly stereotyped, meaning genetically predetermined. The advantage of this mechanism is that you can secure a considerably reproducible "Bauplan" (a blueprint involving distinct morphological features of a body plan) in the development of a system. Firestein referred to the example of the motor system, where nerves are coming out of the spinal cord to connect with designated muscle tissue. But how do these motor neurons know to which muscles to go? They follow chemical gradients that attract them and, after bundling them together, make them give up their nerve bundles to travel out at a specific point. "The idea is that there's a gradient of some chemical, from low concentration to high, and that axons have a receptor on them that's sensitive to this chemical attractant."

It stands to reason that the axons of olfactory sensory neurons do the same, that they developed along a designated chemical trail to their glomeruli. Accordingly, each receptor neuron should grow in the route of a predetermined trajectory from the epithelium to the bulb. Say a neuron is expressing the receptor R1 and is going to one place in the bulb, glomerulus {GR1}. Another neuron is carrying the receptor R2 to a different location, glomerulus {GR2}, and so forth. But a series of subsequent experiments, one after the other, have seriously undermined this premise. Three strikes, in fact, and the notion of the genetically predetermined wiring of the bulb was out.

The first obstacle for a stereotypic wiring of the bulb arrived with another experiment by Mombaerts.[25] Considered a routine inquiry, its results took everyone by surprise. "I was fortunate enough to be right there when they were doing it," Firestein recalled.

"A lot of work we did here was on this whole wiring diagram. Not the wiring diagram of the bulb, but how the whole epithelium *wired to* the bulb. Why are these axons expressing the same receptor all bundled up together into a glomerulus? The standard story in the field was that somehow or another, these receptors were detecting something that helped these axons find their way to a particular glomerulus." Firestein emphasized: "What Peter showed was that that doesn't actually seem to be the case."

Mombaerts's experiment followed a simple plan: select one potentially determinant factor, isolate it, and alter it. Then observe its effect.

Mombaerts's lab engineered three kinds of mice, with green fluorescent proteins (GFPs) attached to specific receptor genes. Fluorescent proteins enable scientists to observe the path of axon development. One mouse population had a GFP connected to receptor gene I7; in another mouse GFP tagged the receptor gene M20; and the third mouse population had the neuron with the M20 gene replaced with the gene for I7 and pinned with a GFP. Where did the modified ex-M20 neurons with the I7-replacement gene go?

"You would have predicted that they go to the I7 glomerulus," Firestein suggested. "Because we know that the whole idea was to prove that the receptor was crucial in getting the axons to the right glomerulus. And the only thing you've changed is the receptor." Contrary to expectation, the neurons did not terminate in the I7 glomerulus. Were the axons targeting the M20 glomerulus instead? "The answer is neither!" Firestein exclaimed. "The answer is [that the modified neurons] make a glomerulus *over here* or *over here* or *over there*." Firestein started drawing various random points on a sketch of the bulb. "It's not even clear what the rules are!"

This was strike one for the stereotypic wiring of the bulb. It was soon followed by strike two. What would happen if you replaced a gene for an odor receptor with a gene coding for a nonolfactory receptor—specifically, a receptor for a member of the same (genetically related) protein family? "The real nail in the coffin," Firestein argued, "a result that almost never gets talked about, is that by Paul Feinstein."[26] Feinstein, a former postdoc in Mombaerts's lab, replaced the I7 odor receptor gene with the gene for the ß2-adrenergic receptor. ß2-adrenergic receptors are adrenaline receptors; they are part of the same protein superfamily, G-protein coupled receptors (GPCRs), and thus similar enough in topology and structure. All GPCRs have seven transmembrane helixes and share a substantial number of amino acid sequences. Yet their function is entirely different. Adrenalin receptors regulate the activity of the sympathetic nervous system. Odor receptors bind odorants.

Feinstein's modified neurons had nowhere to go. They "no longer have an odor receptor that could get them anywhere," Firestein clarified. "There's no way that any mouse ever existed on the face of the planet that had a ß2-adrenergic glomerulus waiting for its axons, especially the ß2-adrenergic receptor, to come to it." Everyone was in for a surprise, again. All neurons carrying the adrenaline receptor gene converged in one glomerulus in the bulb, just as any other neuron with odor receptor genes. "So that shows definitively that there's no map," Firestein modified his statement, "no preexisting map, on the surface of the bulb to which these axons then go. If there is a map, it's an induced map. It's induced by the axons of the cells themselves."

Strike two. These two experiments, swapping odor receptors with other olfactory receptor genes or nonolfactory GPCRs, created axons that had no predetermined place on a genetically prewired map. Last, what would happen if you knocked out the receptor genes in some neurons entirely? Technically, their axons had to go somewhere in the bulb, just without a designated driver. "Well, they don't form a glomerulus," Firestein replied. "They go off all over the place." Was this not an indicator for genetically predetermined wiring? "The initial interpretation of this was: ah, see, [the sensory neurons] have no way of getting to their target because they don't have the receptors anymore. They're just wandering around. They don't know what to do." That interpretation soon fell apart. It turns out that it is impossible to knock out an olfactory receptor gene. When you knock out the gene in an olfactory sensory neuron, the neuron quickly decides to express another receptor gene instead. (Sensory neurons have a range of possibilities from which to pick a replacement gene. Within this range, the choice is random.)

"The interesting thing," Firestein added, "is that these axons don't wander all over the bulb. They stay in a region of the bulb." So if a cell cannot express a receptor gene, it will pick another one. Such choice is not random. A cell won't pick any receptor out of the entire range (one thousand receptor genes in mice) but from a limited number. "It's some group of them that, somehow or another, is controlled together. And therefore all [neurons] go to some region of the bulb. But where they are on the bulb, they don't care." He continued, "this is an experiment that should have been done years ago, knocking different receptors out and seeing how many regions there are in the bulb." What had looked like totally arbitrary axon growth in this knockout experiment was, in fact, neurons driven by multiple genes.

Strike three.

This series of gene replacement studies posed a fundamental problem for the idea of a stereotypically arranged bulb. There is no preexisting arrangement of places to which the axons of the sensory neurons will grow. "The problem with the map business, it's that there's clearly no predetermined map on the surface of the bulb," Firestein confirmed. This is special to the olfactory system, as far as we can tell. "That's not true in other sensory systems so much," he added. What is evident is that receptor genetics constitutes a critical factor. Less obvious now seems to be what mechanisms guide the axons in their development. "Does [a map] form as a result of the way these axons distribute themselves? I mean, *some* kind of a map forms, but is it a functional map?" Firestein paused. "I don't know."

The Map as a Developmental Artifact

"Maybe it's just a developmental map," Firestein proposed. "It's just the easiest way for [the olfactory bulb] to develop." Axons could be guided in their attraction of the receptors toward each other. This hypothesis, Firestein remarked, "first floated in Peter Mombaerts lab, which we adopted as well. It more or less came out of conversations between our labs and, to some extent, with Charlie Greer." The resulting hypothesis was "that there are no glomeruli preexisting on the surface of the olfactory bulb. But the cells are attracted together. They fasciculate [form bundles], and then they travel for a while, and when they reach an open spot on the bulb, they plop down and begin to form a glomerulus. And then, everything else forms around them to become a glomerulus, all the other cells that are involved in making a glomerulus."

Firestein and Zou explored this idea.[27] Looking at the brain development of very young mice, they observed different stages in the maturation of the bulb. These developmental stages did not square with the model of the stereotypic placement of glomeruli. Rather, they found these stages suggested that the spatial arrangement of glomeruli built on sensory experience during maturation.

Two findings of their study stand out. First, glomeruli are not always homogeneous in young animals. They form heterogeneous clusters during early development, meaning these premature glomeruli consist of neurons expressing different receptors. Second, axons did not just grow together, settle, and converge in glomeruli, as previously assumed. Axons first spread out to different placements instead. Glomeruli subsequently were pruned and refined by activity.

The problem was that earlier studies of the bulb had "more often just looked at the mature animals," Zou explained. "Previous work showed that at a very early stage, there are a few glomeruli, maybe a few hundred. Later, you get the number increased to around two thousand per side of the animal, per bulb. So there's a continuous addition of glomeruli." However, Zou saw things differently: "In my mind, glomeruli formation is a continuous process. It's not like everything is already fixed. It's more like a dynamic process. Not everything is set up at the very beginning." Zou had noticed that (contrary to the received view) in very young animals you could observe *more* not fewer glomeruli than in older animals. "I was kind of shocked. At that time, people thought if you see two glomeruli [in the place of one], they're just noise. So I first thought . . . I got something wrong. The staining was not right. But then I saw it over and over. So, I thought: That has to be the real thing."

No chemical trail. No homogeneity in the formation of glomeruli. Multiple glomeruli in the same spot. Together, these observations seriously thwarted prewired sensory topography. These striking developmental insights, though, did not receive broader attention in the olfactory community, which had shifted its focus to computational models of odor coding. In consequence, inadequate notions of stereotypical representation and chemotopy continue to define large parts of contemporary research on the bulb.

Neglect of developmental mechanisms may have two reasons. First, the majority of studies on the bulb are tracing brain organization in adult mice. The bulb's mature design certainly looks stereotypic; the correlation between the structure and the activity here is genuinely striking. The mature bulb seems as if the axons are headed straight toward genetically designated loci. As if, as Firestein noted, "there must have been some chemical that was up there that was attracting the axons." Second, Greer remarked on a general tendency to neglect developmental studies. "Historically, there have not been that many people that have followed [the system] developmentally." Consequently, the premise of the bulb's stereotypic arrangement is seldom questioned. Zou agreed: "A lot of times, the people who did most of the genetic manipulations were not from a neuroscience background. So, they don't care too much about development." A central problem in the field of olfaction has always been its cross-disciplinary fragmentation. Development is critical to understanding the wiring of the bulb, as determinative of its neural representation. The question of what part of a sensory system is hardwired or contingent upon experience during development also played a central role in earlier vision studies. In olfaction, glomerulus formation fundamentally depends on receptor genetics and developmental mechanisms. Zou agreed: "Only if you get to the details will you know that the olfactory system is quite different from other systems."

Granted, the bulb's development shows variation. But the final arrangement of glomeruli seems to display invariant glomerulus placement and reliable patterns of odorant activity. Didn't we see stable patterns, something like a fingerprint for each odorant in the bulb, even across individual members of a species?

A closer look tells that neither location nor functional activity in the mature bulb is genuinely stereotypic either. "If you look carefully," Firestein noted, glomeruli "are not exactly in the same place. They can be five to eight glomerular diameters off in the area in the bulb, just some area in that they're more or less, and that confines it." (Glomerular diameters in mammals are 30 to 200 μm.) This stands in contrast to most of the review literature, which commonly labels the bulb as

"stereotyped," "constant," "topographically fixed," "topographically defined," "topographically stereotypical," "stereotypically positioned," "with precise stereotypy," "invariant," "spatially invariant," "precise," and "nearly identical."[28]

In 2015, Mombaerts broached the subject again, asking: "Exactly how precise are these precise positions?"[29] This question had received surprisingly little attention. Mombaerts's lab found that the answer depended on the glomeruli. Variability in glomerulus position was higher than usually reported. Further, in testing six different types of glomeruli, researchers found that variation was not homogeneous. Some glomeruli showed more variation in their location than others. One way to make sense of this is to see whether variation in glomerulus placement relates in any way to the overall tuning range of the different receptors expressed in the sensory nerves. Receptor genetics continues to put an end to the idea of a generic spatial Bauplan of the bulb.

Greer was quick to highlight another challenge to understanding odor coding via stereotypic function: the bulb's heterogeneous activity. Electrophysiological recordings of activations in the mature bulb showed considerable variability at the local circuitry (which we explored in detail earlier). "If you take a look at any given glomerulus, there are anywhere between eight and maybe twelve mitral cells that inhabit a particular glomerulus. They do not have the same response patterns. So, you present an odor, and you are recording from two mitral cells that inhabit the same glomerulus. One may respond, the other one may not. Or the temporal aspects of the response patterns may be different as well."

Electrophysiological research on the circuitry yields a considerably more heterogeneous picture of activity in the bulb than does staining and neuroimaging of glomeruli. Several factors of variation are critical in this context—two in particular.

First, activity in the bulb varies between awake and anesthetized animals. In addition to the evident difference in sniffing behavior, neural responses in anesthetized animals exclude afferent feedback and top-down factors such as attention, motivation, and previous experience. Such studies, therefore, rest on an artificial model of linear sensory processing from lower to higher sensory areas. The profound impact of prior experience on unit activity in the bulb is not negligible. Nathalie Buonviso et al. demonstrated that the cells of rats under prolonged odor exposure (twenty minutes) showed a decrease in activity twenty-four hours later.[30] Additionally, activity in untrained rats varied daily, as it did with motivational factors.[31] In other words, experience and training, motivation, and other decision-making factors determine neural activity. (The perceptual implications of this are explored in Chapter 9.)

The second factor concerns divergence in single cell and cell population recordings. Single cell recordings, often in anesthetized animals, cannot sufficiently account for the functional characteristics of odor coding. Besides, population coding may be more determinative of odor coding (Chapter 8). The upshot of this detour into electrophysiology is that our understanding of the bulb, in particular the strong emphasis on its topographic activity, is far from settled.

Also, it is likely that the relative stability and relative invariance in glomerular organization and patterns do not occur in genetically diverse organisms and species, or in individuals across a species whose brains are shaped by multiple experiences outside the highly controlled life of caged lab mice. The appearance of a bulbar map may be enforced by experimental artifacts, based on methodological designs or the material regulation of laboratory standards.

Notably, our understanding of the olfactory system builds on genetically homogeneous model organisms like mice, rats, and fruit flies.[32] What would happen if we were to look at genetically heterogeneous organisms, like humans? Christian Margot pointed at a study by Charlie Greer.[33] This work "showed that little had been known about humans because part of [this knowledge came] from cadaver analysis, older people who donated their body. Charlie Greer analyzed the glomerular [material] of young people and found something like 5,500 glomeruli." This finding was unexpected. Previous estimates had built on a conversion ratio of 2:1 (of 350 receptor genes to 700 glomeruli). Greer's study now suggested a ratio of ~16:1! "So clearly there's something very different. Nobody has an explanation for this," although such a vast number of glomeruli may link to the robustness of olfactory signal representation.

In the end, the periphery of the bulb and its activity patterns is less stable and invariant than it appears. Instead of a prewired Bauplan, the bulb yields a developmentally instructed representation of receptor activity. Firestein concluded: "Almost anything you do, genetically, to the animal, that affects receptor expression, changes the glomerular organization of the bulb. My feeling is that the glomerular organization of the bulb is the most plastic, malleable, easy to disrupt thing there is. It's not stereotypic. It's not hardwired. It changes around."

Stimulus Representation, Beyond the Map

The wiring of the bulb is not predetermined, and receptor genetics plays a critical role in glomerulus formation. What appears disordered is not without function. The olfactory bulb does not provide an odor map if understood as a chemotopic or rhinotopic organization, or some stereotypic odotopy based on stimulus chemistry. So, stimulus topology is not what is encoded in the bulb. But the central puzzle remains: What does such a coarsely structured, developmentally induced arrangement communicate?

"The problem is you don't know exactly what to look for," Reed replied. "We don't have any good models for how to address these questions. What I find interesting is that, to some extent, these are questions or problems that don't exist in the visual and auditory system. They are not correlates of this kind of problem. The escape from that is either to say that there is some sort of 'odotopic' map in the bulb, which is essentially trying to cram it onto the two models that we know about. Or you have to say that it uses a fundamentally different process."

The solution, Shepherd thought, lies in the computational mechanisms that underpin olfactory perception. These processes may not be fundamentally different from vision but they build on a mechanism of feature coding in the visual system that is likewise the least understood: face recognition. "The new technology of facial recognition, I think, has great implications for eventually a precise quantitative characterization of the irregular olfactory glomerular patterns—another example of how olfaction can benefit from mainstream science and technology."

Faces, like odors, are not defined by simple structures like shape and general classes of forms (or we'd all be artists). What makes faces so mesmerizing is that we recognize them by their individuality among general patterns. Shepherd's proposal is appealing because of the decisive feature of odor perception: its distinctiveness, especially in context. Shepherd was convinced that the analogy with facial recognition would make sense of activity in the bulb: "That prediction will be realized by the new technology of facial recognition, when it, or something like it, is applied to glomerular active pattern recognition." Shepherd's computational comparison may turn out right without having to save the notion of topography.

The unpredictability of the stimulus, including the irregularity of chemical topology in odor recognition, need not imply a lack of regularity that cannot be modeled. It just necessitates a different modeling outlook. We need to rethink what is computed and, in turn, represented.

What characterizes olfactory encounters "in the wild" is the unpredictability of the chemical stimulus in its environment and its interaction with the sensory system. In reply, Thomas Cleland, at Cornell's Department of Psychology, suggested that what the bulb represents are not chemical classes, but the chemical environment. More precisely, the bulb tracks the statistics of a changing odor environment.[34] Activity in the bulb may express environmental statistics of the chemical environment.

Such a *functional topology* requires a sufficient degree of flexibility and plasticity in a system's wiring, not anatomical proximity (like in vision). Specifically, it requires a setup optimized for learning about input frequency and changeability. This setup builds on two critical features of the olfactory system: receptor genetics and inhibition.

Receptor genetics is critical to the statistical tracking of the chemical environment via the tuning and tuning range of receptor types. (Recall that receptor genetics link to ligand affinity in receptor binding.) This may explain the coarse overlap in activity patterns to specific odorants. Instead of stimulus topology, the spatial organization of the bulb expresses a coarse representation of the tuning range.

Inhibitory processing via the delicate microcircuitry in the bulb, in this context, serves as contrast enhancement and regulates the discrimination of similar stimuli. Local processing of odor signals in the microcircuit machinery of interneurons does not merely coordinate and refine spatial patterns. Notably, it induces a temporal signature that demarcates specific olfactory signals and further distinguishes them, especially in comparison to chemically similar odorants or different stimuli with overlapping activity patterns. How the principle of temporal coding undergirds odor recognition and categorization, therefore, will be the subject of Chapter 8.

From a systems theoretical account, we can here conclude that the neural representation of odors is not defined by chemical topology but constitutes an expression of environmental classes. After all, Terry Acree quipped: "The brain is not a German chemist!" (Shepherd quickly retorted: "But the brain of a German chemist is.") Neural representation hinges on the interactions of the organism with the features of its environment. That also indicates the need to model those features on the conditions afforded by the system. Consequently, and following this chapter's receptor-centric view of how the system gets wired up, the next question must address central processing: How does the brain learn to keep track of and memorize environmental regularities in stimulus interaction?

Beyond Mapping, to Measuring Smells

Consider that your brain is living in perpetual tension with itself. Its "wants" and its "needs" seem to diverge. Like a sniffer dog, it continually seeks out information. Driven by the search for relevant or new bits of information in the environment, your brain is always adjusting and reacting to your changing physical and mental states. (The world can look different to you when you are hungry or grumpy.) At the same time, your brain wants stability. It wants to know what it is doing, or, more accurately, what kind of order underlies all this signaling chaos from the outside world that hits it on an ongoing basis. The brain copes with this plethora of data by making predictions based on anticipated regularities. Here, top-down effects (informed by previous experience) constitute a central condition for learning and memorizing specific sensory features to assign them perceptual categories.[1]

John McGann sounded convinced that this is how the brain works. "The brain is making smart guesses about what's out there. It's just peeking at the input every once in a while to see if it's true. I think that there may literally not be an independent truthful purely bottom-up signal to look at. It is a model of: Here's the guess. Does it match? Let's adjust if it doesn't."

The idea of the brain as a forecasting machine is not new. Herrmann von Helmholtz proposed similar ideas in the nineteenth century. Erich von Holst and Horst Mittelstaedt, in parallel with Roger Sperry, described the predictive effect of "efference copy" (or corollary discharge) in vision in 1950.[2]

A simple test demonstrates the idea. Hold your index finger in front of you and look at it. Start moving your finger sideways: left and right, and left and right, and so on. Track the moving finger with your eyes. At a certain speed, it is harder to keep track of the finger with your eyes; your visual image of the moving finger blurs. Now hold your finger still while you start shaking your head left and right, and left and right. Fix the finger with your eyes. This time, your visual image of the finger does not blur but remains relatively stable.

What happens is that your brain makes an internal copy of the anticipated motor behavior required by your visual system. This involves two predictive processes: First, when your eyes track the moving finger, your retina follows the movement of an external object (exafference). At some point, the retinal motion will be too slow to keep up with the speed of your finger and, as a result, the visual image of the finger blurs. By contrast, when you shake your head while you fix your eyes on the finger, your brain anticipates the appropriate motion of the eye to fixate the resting finger (reafference). As a result, your retinal movement is at speed with your head movement in the perception process. This memory function couples the motor system with sensory regularities—an educated guess of where your eyes ought to be (instead of where they are). Terry Acree nodded: "Remember the brain evolved from the body and not the other way around."

Forward models enable the brain to send motor signals to the eyes and compensate for your movement in relation to a static environment. It is a tradeoff between accuracy and speed. By creating a simpler model (it does not need to be accurate, just "good enough"), your brain reduces the amount of signaling that it has to cope with. Self-induced signals are distinguished from signals of the external input. Your brain reduces its own "cognitive load" by decreasing stimulus noise and stabilizing the processing of incoming information via self-generated signals. (This mechanism is not unique to humans: flies, fish, cockroaches, and crickets do it.)

Recent work in cognitive neuroscience and philosophy jumped on the idea. Forecasting mechanisms were mainly studied in vision, but also in audition. Meanwhile, most models approach the predictive brain from a computational angle with insufficient connection to the cellular machinery in wet-lab research. Olfaction could bridge this gap.[3]

An Open, Chaotic System

Systematic studies on the predictive powers of the nose are meager. An early exception to this was Walter Freeman at Berkeley. In the 1980s, he advocated a nonlinear dynamic model of olfactory processing that resonates with modern ideas about the predictive brain.[4] Freeman assumed that brain regions such as the bulb adopt a set of steady states, an equilibrium of neural activity, gravitating toward "attractors" as preferred states. In olfaction, exhalation and inhalation determined these steady states in their receptiveness toward information from the external stimulus (during late exhalation and early inhalation) or information integration in domain-internal neural activity (exhalation). Against this constant background activity, odor stimuli could be recognized as familiar or learned as new. Known odors elicit an established spatiotemporal signature of activity. Unknown odors first evoke chaotic activity before acquiring their own spatiotemporal signature for future recall. Chaos here was a condition for learning so that the brain would not confuse a novel odor with the signature of an already known one.

Freeman's model was ahead of its time with its implications for the neural representation of odor. He suggested that the activity of the entire bulb was involved in odor recognition. Spatiotemporal patterns were not "representations of odors" because the same odor could acquire different activity patterns. These patterns were contingent on not just the specific background state of the animal (Freeman worked with rabbits), but also on the behavior associated with an olfactory stimulus. The cognitive scientist Antony Chemero gave an example: "That is, if a rabbit has learned to associate a particular behavior (say, salivating) with a particular odor (say, carrots), it will reliably produce a characteristic pattern of activation across its olfactory bulb. But if the rabbit is taught a different behavior (say, cowering) to go along with the odor of carrots, it will then produce a different pattern of activation. So, [Freeman and his colleague Christine Skarda] claim, the pattern of activation cannot be a representation of 'carrots.'"[5]

Freeman rejected the idea of representation as neural correlates of fixed schema or templates. He understood neural signatures as how the brain learns associations between a stimulus and a response. The same stimulus may be involved in multiple signature patterns according to decision-making context (being involved in different behaviors, or combined with cross-modal cues).

Leslie Kay, a former grad student of Freeman, remarked on the idea of nonlinear dynamics: "It certainly is a good metaphor. The system acts as if it's a chaotic system." However, "it's not a closed system, so you can't ever prove it." Still, Kay remarked, you can study its effects: "Walter showed that if you change the association of the odor, the pattern changes. Not only *that* pattern changes, but the other patterns for the other odors also change. Every time you learn something, it rearranges the bulbar network." She found these effects in her research, too.[6] "We were using behavior to induce the oscillations or brain activity that we want to study. If we tweaked the task a little bit, to make it easier to train the animal, the way the brain did the thing would change. The sequence is the same, and the relative timing is more or less the same. It's predictable. But the amplitude of these things is all over the place, and it's significantly driven by the task: how many tasks the animal knows, which odors they're sniffing, where they are in their learning, whether they're learning the first odor sets or subsequent odor sets."

Olfaction is a task-driven system exposed to a deeply irregular stimulus, unpredictable both in its environmental occurrence and structural permutations. How your brain allows for flexible behavior that deals with such a chaotic world has a bearing on how to conceive of the idea of forecasting in recent cognitive neuroscience. So what can olfaction reveal about the concrete, cellular mechanisms that undergird the predictive brain?

Measuring the World

Imagine for a moment that you experience the world as your brain does. Like the brain, you cannot know the identity of a smell beyond what is happening in the nose. All you can "see" (for lack of a better term) are the signals from the receptor sheet and the neural activity over several processing stages. *When, how fast,* and *in what order* a signal occurs might be more fundamental than *where* it occurs. That invites a change of perspective. Instead of fixating on odors as firm templates, we should consider the neural principles by which olfactory signals are computed into smells, categorized into more or less regular perceptual patterns. Olfactory processing, the previous chapters have shown, is an evaluation of frequency in familiarity and novelty of unpredictable, changing information in our chemical environment. This chapter argues that it takes the brain to be something like a measuring machine to deal with smells.

What does your brain do when making predictions? It measures the environment in its various dimensions in relation to yourself. For example, how close or far are you from something? Does it move, and does it move fast or slowly? Does it change its quality over time, or does it stay relatively stable? Does it appear to be harmful or pleasurable? and so forth. These are not properties intrinsic to objects and environments per se; rather, they constitute judgments about the input concerning the perceiving organism (Chapters 3–5). The mechanisms of mixture coding thus require a flexible integration and classification of irregular stimulus signals into variable behavioral contexts.

This idea does not square with stereotypic, topographic maps where the meaning of an odor is represented by fixed neural representation. Induced by receptor behavior, neural activity encoding mixtures does not furnish us with transparent stimulus-response maps. The olfactory stimulus cannot be captured on an additive scale since its coding and computation is not additive—not in the bulb and even less so in the olfactory cortex.

Alternatively, the critical hypothesis here is that olfactory signaling constitutes a measure of changing signal ratios in an environment informed by expectancy effects from top-down processes. As ratios (an idea first encountered in Chapter 6), smells are an interpretation of signal combinations and magnitudes.

To measure its environment, the brain requires a way to structure and calibrate incoming signals. That can involve multiple kinds of mechanisms spanning several processing stages. At the periphery, we saw that selective adaptation acts as calibration of the system to evaluate changes in the chemical environment against some form of background. At the level of central processing, the measuring activity of the brain does not necessitate a spatial map, although it may develop and make use of one for certain tasks or in appropriate instances. In olfaction, we find that stimulus information is first broken down into several bits and pieces, which are then combined and weighed against each other to form an overall sensory impression. The precise mechanisms in central processing that structure and govern this wide distribution and integration of olfactory signals into multiple parallel processes move into focus next.

Odor coding has a deeply dynamic format. The olfactory system can assign multiple meanings to the same stimulus. In addition to Freeman, the neuroscientist Gilles Laurent at the Max Planck Institute in Frankfurt considered this the critical feature to model odor coding, arguing in favor of network models instead of discrete chemotopic representations. To encode multidimensional signals under varying decision-making contexts, the system employs two principal processes: the formation of an extended coding space, and the wide distribution of neural signals.[7]

First, the coding space in olfaction indeed is sizeable. Previous chapters showed that the chemical stimulus is (1) first broken apart via combinatorial coding at the receptor sheet (Chapter 6), and (2) further decorrelated via the inhibitory / excitatory dynamics in the microcircuitry of the bulb (Chapter 7). Decorrelation is essentially a temporal process, built by local processing in the molecular machinery of interneurons. Each decorrelated olfactory signal manifests a temporal signature that allows it to pair with other signals in similar, synchronous states. (We will get to the cellular mechanisms that undergird this process soon.) The spatiotemporal activity in the bulb should thus be seen as an expression of the dynamic coding space—not a fixed representation of odors, since odorants can be assigned various meanings and, in turn, patterns.

Second, decorrelated signals then are widely distributed and sparsened in the olfactory cortex. Their wide distribution allows olfactory signals to be integrated and synchronized with parallel processes in neighboring cortical domains (allowing for timely combinations with cross-modal, verbal, and other cues). Sparse coding, in this context, reduces overlap in the complex neural patterns of a multidimensional stimulus. Specifically, sparse patterns allow the fast and also transient formation of manifold associations of specific olfactory signals with distinct meanings and varying conditioned behavioral responses (like in Freeman's rabbits). If your signals were too detailed on a neural level, their linking with other signals would become very complicated in connectivity—and a chore in later recall. Sparse codes may be less detailed, but faster in processing and recognition (a hypothesis that notably resonates with the Laing limit, in which the analysis of complex odor images as "Gestalts" is limited in component resolution; see Chapters 6 and 9).

The key to the dynamic computation of odor images resides in temporal processing, not topographic patterns. Measurement is thus a more apt metaphor than the idea of a mapped projection for how the brain makes sense of fluctuating information pieces.

The idea of measurement also changes the notion of representation. The brain does not yield fixed patterns for objects (pattern X stands for odor X) but dynamically encoded signatures that link input information with other responses (signature X has been formed to indicate perceptual-response state X).

What kind of neural architecture and cellular mechanisms allow the olfactory brain to receive input and operate in this manner, to sample and measure than represent and map the variable composition of its chemical environment? The extensive connectivity of bulb and cortex with neighboring cortical regions provides insight. The olfactory cortex is a collection of several domains. A large proportion of mitral cells connect the bulb to the piriform cortex, the biggest region of the primary olfactory cortex. Its architecture facilitates a wide distribution of odor signals.

The Enigma Machine

Entering the olfactory cortex is like walking into a vastly distributed neural firework of combinatorial activity. "What is the name of the machine that was used by the Germans?" Acree laughed. "The Enigma! You are running the input through an Enigma machine. The information is still there, but it is unbiased by the input. You can't tell from the structure of the input what the information is, or from the output what the structure of the information is. That's why the Enigma was a great coding device, but it had to have a mechanism for decoding it." He looked thrilled. "In a sense, the connection between the olfactory bulb and the piriform cortex is an Enigma machine." The Enigma was solved not via a descriptive account of its output, nor by having the code for each message. There was no apparent match between input and output. To crack the Enigma required a decoding of the encryption principles enacted by its machinery. The same idea applies to the olfactory brain and how it translates molecular bits into mental images via their ratios.

The piriform cortex is a busy information hub in its output connectivity. It sends its signals to a number of neighboring cortical domains involved in all kinds of cross-modal or cognitive functions, including the hippocampus (memory), amygdala (salience, affect, and avoidance), olfactory tubercle (cross-modal, auditory associations), orbitofrontal cortex (decision-making), entorhinal cortex (navigation and perception in time), perirhinal cortex (cross-modal integration), and more. The piriform does not just send projections to these areas; it also receives robust feedback from its neighbors in return.

The olfactory cortex, as a coding device, scrambles information. Instead of forming discrete neural clusters, like the glomerular layer, the axons of the mitral cells projecting from the bulb spread out far, wide, and randomly into five major cortical regions of the olfactory cortex and neighboring areas. "Some of the regions maintain some spatial order, like the amygdala," Richard Axel explained. Others do not. "If you look at the primary sensory cortex, you see that this order is gone. Now all odors activate a distributive representation, which is interdigitating and different for every odor. A given odor will activate a different representation in different brains . . . and in the two sides of your own brain!" So your piriform pattern responding to 2-methoxybenzaldehyde (with a sweet, anise odor) is not just different from that of your lover, its activity in your own right hemisphere even looks different from that in your left hemisphere.

The piriform was considered the critical area of odor object formation, where the majority of signals from different olfactory regions are integrated into the neural correlates of smell percepts. But the piriform yields no apparent order.[8] In 2011, Dara Sosulski, together with Richard Axel and Sandeep Robert Datta (previously at Columbia, now Harvard), followed the projections of mitral cells from the bulb into the piriform.[9] They injected mitral cells innervating a glomerulus with a neural tracer (TMP-dextran) to find that axons signal to their closest neighbors, as well as to other axons further away. At the cortical level, we face a display of widely and irregularly distributed information (Figure 8.1).[10]

It's baffling to see this neat clustering in the bulb discarded in the sensory cortex. Leslie Vosshall, once in Axel's lab, said: "I've felt people are nervous about Axel's piriform cortex work because it's hard to prove that something is random and flexible. But I found that work to be a huge relief because I didn't even feel like we were going to get anywhere if every synapse you look up it's a one-to-one mapping. At some point, you have to abstract meaning from these neural patterns. It has to be flexible so that you can have a percept of something that you've never smelled before. You may not be able to put a word to it, but you can detect it. It has to be layered with experience. So I found the flexible coding in the piriform cortex a fantastic result!"

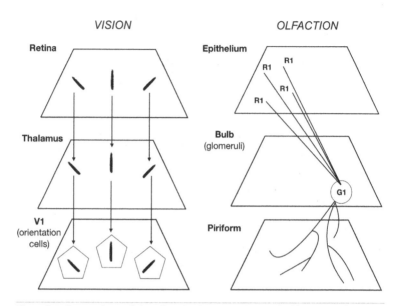

Figure 8-1 Comparison of signal projections in vision and olfaction. Left: retinotopic organization, showing the topographic projection of signals from retina, via the thalamus, to the primary visual cortex (V1). Right: spatially distributed receptor activation in the epithelium layer; the activation signal is collected and organized into discrete spatial clusters (glomeruli) in the olfactory bulb before being dispersed into different areas of the piriform cortex. Source: © Ann-Sophie Barwich.

Rather than molecules, your brain depicts transient information patterns, extracted and weighed in a given context, without a superimposed matrix of chemical classes to accommodate for countless molecular permutations.

Matt Rogers, however, cautioned that some kind of map might reveal itself with better tools. "We only see 10 percent, 15 percent, maximum 20 percent" of the piriform. "We do not see everything with the tools we have. Until someone can show me the entire piriform, a full 3D view, I'll leave the idea open that this chemotopic map doesn't go beyond the olfactory bulb." He added: "I know there's no evidence of that now, or very little. On the record, I don't think there is a map. I'm just raising the bigger question. We try to get the best information with the tools and technologies that we have."

Avery Gilbert remarked that even some hidden map seemed insufficient to account for odor coding as long as it is modeled on the properties of the visual system. Olfaction, previous chapters have shown, does not fit that schema. "I think olfaction is not going to be like the cortical mapping of vision in all its ways. The color stream areas, the edge detectors, the color constancy—you can locate all of that stuff. That's at the hidden plumbing level of the system, completely at a different level. And I think we've done a good job of finding the biologically, neurologically invariant parts of vision. I think the invariant parts of smell are all very peripheral."

Besides, Axel noted, the visual system is not reducible to topography either. Higher-level vision abandons topography. "The notion that you are continuing to use topography as you solve a problem, say, of object recognition—this is a pretty simple problem compared to what vision really has to do. You're going to discard spatial order and use random inputs." Your brain cannot possibly represent all the objects it encounters. At some point, it operates on a more abstract level. "I mean there is structure, but think vision is ultimately going to come down to random inputs again." He added: "This idea of randomness is going to be seen in later states in the other senses. My feeling is that other areas that have this beautiful topographic organization soon discarded it as one moves to increasingly high brain stations."

Olfaction becomes an ideal model for higher-order processing in other modalities. Higher-brain integration is notoriously tricky to understand; its signaling is nontopographic, seemingly random, and autoassociative, just like in the olfactory system.

Recent models describe the piriform cortex as autoassociative architecture. (This includes work by Lewis Haberly, Don Wilson, Kevin Franks, Naoshige Uchida, Philippe Litaudon, and Jeffry Isaacson, among others.)[11] Autoassociative refers to the interactive and cross-connected processing of the piriform with neighboring areas. Most of the piriform's activity is characterized by feedback, which sharpens, inhibits, or enhances the information. This includes feedback processes from the piriform back to the bulb as well as feedback loops between the piriform and higher cortical areas. Take a look at Figure 8.2; what happens between the bulb and the piriform is most vividly described as "scrambling the signal," *again*—this time without recovery. What you see is that the projections of some glomeruli concentrate in some regions of the piriform cortex (signal convergence). Meanwhile, the projections of other glomeruli stray and scatter into entirely different areas (signal divergence).[12] An open question is whether this may relate to genetic factors, perhaps the diverse tuning range of receptors. Axel explained the implications of such interdigitating connectivity: "Experiments of others, as well as experiments of our lab, showed that a given neuron in primary sensory cortex gets inputs from several of these glomeruli. But the inputs it gets are also without order. A given neuron in the sensory cortex gets input from a random combination of glomeruli."

"If that's the case," Axel emphasized, "then a given odor must activate different representations in different individuals, and it implies that these representations have no [intrinsic] value; they can have no inherent meaning to the organism. Now we need higher brain connections to impose meaning on these representations. We did a lot of experiments which seem to indicate that this puzzling observation is indeed correct."

Input activity in the piriform is distributed and scattered; it is relative to individuals and temporal factors. And so Vosshall said,

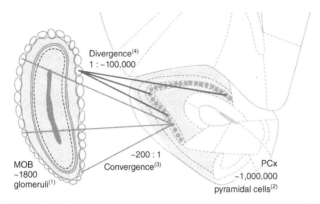

Figure 8-2 Signal scrambling between the main olfactory bulb (MOB) and the piriform cortex (PCx) illustrating signal convergence and divergence. Source: M. I. Vicente and Z. F. Mainen, "Convergences in the Piriform Cortex," *Neuron* 70, no. 1 (2011): 1–2, fig. 1 © 2011 Elsevier Inc. All rights reserved.

"I like the idea that the piriform cortex is this erasable read-write logic machine that will display information that has no fixed location or identity. But I don't know how that's read; I don't think anyone knows how that's read. But it seems important to me that you don't just . . . the worst possible thing is that you take the four hundred glomeruli, and then you have some geometric arrangement where the four hundred glomeruli map onto piriform cortex to be a spiral or a cube because that doesn't solve the problem of perception. There's not enough information coding if you start to impose patterns."

This randomness is functional. Information is broken into countless pieces to be weighted and grouped by various permutations. This is how the olfactory system facilitates striking behavioral flexibility in response to a stimulus that affords different meanings in various environmental combinations. Two mechanisms govern the Enigma machine to impose meaning: top-down processes forming expectancy effects and temporal coding structuring input ratios.

An Output Map

Maybe it's time to reverse engineer. Dong-Jing Zou suggested looking at the output patterns, rather than the input patterns, in this context. "What I'm interested in is the output of the piriform. The output of the piriform probably is not as disorganized as people would think. There may be some organization, for example, like a new type of convergence."

In 2014, Stuart Firestein and Fred Chen examined the piriform by reversing the problem.[13] Instead of asking what the signal is for (that is, by tracing its *efferent* connections), they looked at where the signal comes from (that is, they traced its *afferent* connections). Specifically, they followed projections from two higher-level domains in the orbitofrontal cortex (the agranular insula and lateral orbitofrontal cortex) back to piriform. They got lucky. Firestein's lab found two distinct neural populations with a largely nonoverlapping topographic arrangement. This suggests top-down organization: signals in the piriform are regulated by its feedback connectivity from higher cortical domains.

Zou explained: "These cells are intermingled, but they project to different regions. That starts to suggest there's some level of reorganization from the point of view of the piriform output." The results are preliminary, though. A "problem is that the piriform is huge. There are many other regions to which the piriform will project. That has not been studied carefully."

In one way or another, top-down effects determine olfactory signal integration; signals in the piriform are structured by output. A 2017 study by Alexander Fleischmann's lab, tracing neural subnetworks and temporal patterns in the piriform, strengthened this view.[14] These studies remain preliminary, yet they point us already toward a new model.

Connectivity is key. The deep associative network in the piriform allows for broad distributions and instantaneous integration of olfactory signals with neighboring processes. What an olfactory signal means may depend less on its input organization and more on its participation in parallel processes. At the heart of the olfactory cortex, where countless combinations, connections, and cross-activations are taking place at every instance, we see that neural representation, fundamentally, is a process. The piriform provides a measure of flexible signaling activity, informed by activity from top-down connections. From the brain's point of view, and in the absence of a distinct spatial order, its chief task involves a calculation of the presence and strength, concurrence and sequence of signals. Top-down processes determine signal integration, linking odor categorization to experience, expectancy, and cross-modal effects.

This is half of the story of how the brain measures its environment by selecting and prioritizing information. The other half concerns the regulation of input. The order of appearance, the temporal sequence of signaling patterns, is determinate of input concurrence and integration with other signals to weigh it in terms of salience, familiarity, or novelty. Two mechanisms govern the temporal coding of odor signals: primacy coding at the receptor level and population coding at the neural level.

Primacy Coding

Primacy in receptor coding assumes that odor identity depends on selective receptor activation. Only a small set of receptors—not all—respond first to a stimulus in a given concentration, determining what input the brain receives from the epithelium. Explanations of odor via temporal coding originated in studies of latency effects, also in olfaction, by John Hopfield in the mid-1990s.[15] (Latency refers to the interval between stimulus exposure and response.) Dima Rinberg, a neuroscientist with a physics background at NYU, developed this idea into a computational receptor model.[16] "The primacy coding idea came in our lab a couple of years ago," he said. "It's actually a super simple idea—you can order the receptors by their sensitivity: the more sensitive, less sensitive, and very weak receptors. Basically, what the brain sees."

Receptor patterns change with increased concentration. Rinberg described hypothetical receptor codes for different concentrations of the same odor. An odor in low concentration, for example, excites a small group of receptors {R1; R2}. When this odor is administered in medium concentration, further receptors are excited, forming an extended group {R1; R2; R3; R4}. And additional receptors respond to this odor in high concentration {R1; R2; R3; R4; R5; R6}.

Mixture recognition is a natural fit with this model, Rinberg explained. Some receptors can be suppressed when the epithelium is exposed to a combination of odorants, altering the primacy set for mixtures as a result. So if you present the epithelium with a mix AB and look at the primacy set of active receptors {R1; R3; R4; R5; R6}, it would differ from the primacy sets for the individual odorants A {R1; R3; R5; R7} and B {R2; R4; R6}.

Diverging identities of odor signals, on this account, are mirrored by differences in primacy sets. These primacy sets differ for monomolecular odors and mixtures, as well as stimuli in different concentrations. Configural effects in mixture perception, like Acree's "potato chips" image, find a material explanation; specific ratios activate specific primacy sets. The individual odorants (methanethiol; methional; 2-ethyl-3,5-dimethylpyrazine) exhibit a different primacy code than their blended mixture. Primacy codes further differ for the different ratios of components in such blend.

Rogers found this idea plausible from a systems theoretical view. "The question about temporal primacy that Dima brought to the forefront is really interesting. The assumption being that the most sensitive receptor for a particular molecule is one that is more directly correlated with perceptual outcome." New questions arise about odor coding. Rogers remarked: "In the case of primacy, where you might have twenty or so different receptors, and maybe you have a 'best one'—what is the system doing with the other nineteen receptors? Why are they there? What do they do?" Rogers reminded us of Firestein's suppression effects (Chapter 6). "There was this antagonism bit that changed the code. The extent to which this happens is remarkable. My theory would be that there's some redundancy built into the system. So that if your best receptor is not active—for whatever reason, maybe because you have some antagonists in a mixture—then you can still perceive the odor to a degree. Particularly if this odor is behaviorally relevant." The idea is that "multiple channels of receptor activity [can lead] to the same perception. That's what I mean by redundancy. Receptor activity with respect to mixture encoding is a big question."

Rinberg replied to clarify that his model is purely computational. It does not hinge on the biological properties of its components: "The beauty of the primacy coding model is that it does not depend on all these factors." This view is not universally shared. Kay argued that receptor behavior, including primacy coding, is influenced by the biological conditions of the organism. She added that motor behavior determines the sniffing rate and, in turn, temporal coding. The limit of Rinberg's data from head-fixed rodents, Kay explained, is that "they're lacking the vestibular effects that you get from sniffing." (Vestibular effects link to the control of balance and movement.) She explained, "It looks like there's a preference for slower sniffing in the head-fixed rats. If the sniffing is happening at a slower rate, it means they're extending the stimulus—which makes it easier to perceive, but I'll bet it makes it harder for them to do something else. So maybe they're not filtering out chemicals." The modulation of sniffing rate, fast and slow, mirrors a trade-off between a more detailed, accurate detection and quick motor responses.

In response, Rinberg noted that other studies found that odor recognition does not seem to depend on sniffing rate.[17] This is in addition to Rinberg's 2018 paper on "Sniff Invariant Odor

Coding."[18] Still, such stability could link to other factors, including stimulus memory via pattern recognition. The boundary between odor recognition and memory is under constant negotiation. "It's hard to separate olfactory perception from learning memory," Zou noted. "More often if you apply an odor just once, the animal will memorize it right away. How many repetitive odors can you give before the cells memorize?"

The issue of the (in)dependence on biological factors in primacy coding, or their extent, invites further investigation. Whatever the outcome, the idea of primacy coding is breaking new grounds in odor coding. Its principle affords the sensory system quick interactions with its environment. It also allows for scaffolding in perceptual accuracy. Longer, repeated, or more intense sniffing, in contrast with quick detection sniffs, changes at least the chemical sample reaching the nose, prior to judgment. Such judgment hinges on temporal coding in neural populations.

Population Coding

The brain measures neural spikes, nerve impulses. Spiking activity shows distinct rates and sequences next to temporal patterns in neural populations. These spikes represent when, and in what order, combination, and magnitude, a stimulus is translated into a neural signal. Signaling events never occur in isolation; they always form sequences. They are determined by preceding and parallel signals to determine subsequent ones. Just like a kilo feels lighter after carrying five kilos and heavier after carrying one hundred grams, the measure of a neural signal, including its strength, relates to other signals. The experience of weight is a measure of your body in context, as is the neural response to a stimulus. This analogy holds for odor.

How do we measure temporal coding via neural spikes? Rogers laughed: "This is where electrophysiology comes back into the game!" Electrophysiology records action potentials. When a neuron fires, a complex cascade of ions passes the cell membrane, resulting in changes in voltage and electrical current in nerve cells. Temporal coding of neural populations in olfaction is understudied in mammals. It is more prominent in insect olfaction. A reason that temporal models in insect olfaction have yet not found entry to mammalian work is that rodent and insect researchers form different communities. Prominent work on insects has come from, in addition to Gilles Laurent, the labs of John Hildebrandt, Jürgen Boeckh, and John Carlson, including others, such as Mark Stopfer.[19] Mammalian and insect olfaction, despite significant differences, is strikingly similar in pathway and principles.[20] What can we learn on the fly?

Temporal coding in insect olfaction is regulated by inhibition.[21] Stopfer laughed: "I think inhibition is everything!" In insects, "there's already some evidence of temporal structure when you look at the olfactory receptor cells, which in the insect provide input to the projection neurons in the antenna lobe. These projection neurons show much more complicated patterns of activity than the receptor cells. When you look at the actual firing patterns of the projection neurons, when you present different odors to the antenna, it just jumps out at you."

Inhibition mechanisms structure the signal in each neuron like Morse code. Stopfer continued: "These are complicated and reliable timing patterns. You see patterns of inhibition, followed by excitation, followed by inhibition. You see a burst of spikes followed by inhibition, followed by another burst of spikes, followed by another round of inhibition, followed by another burst of spikes. And it's there every single time!"

There are several temporal scales induced by varying spiking rates of sensory neurons in response to different stimuli. These include, for example, different transient bursts of spikes, tonic excitations (slow and graded) that last longer than stimulus exposure, or combinations of each. Then there are varying latencies, durations, and intensities of spikes. Next to excitation, neural signals are regulated by slow- and fast-acting mechanisms of inhibition, etcetera. Kay added: "So what inhibition does is it coordinates and sculpts the activity and times it. Without that, you just have a mass of input—and when you get to the next cortical area, everything's going to get smeared out."

Inhibitory activities effectively regulate oscillations of neural populations, Stopfer explained. "We see inhibition on two time scales in the antennal lobe, for example. You have these very slow patterns that are sculpting the patterns of the projecting neurons. Superimposed onto that is this very fast, rhythmic oscillation coming through and that's tending to synchronize the spiking. What could this possibly mean to the rest of the system? The idea that Gilles came up with, which I think is correct, is that the cells that are following from them are very, very sensitive to synchronization."

Stopfer emphasized the importance of inhibition in the synchronization of activity. "The idea behind it is that rhythmic inhibition (the oscillations) coordinates the spiking of populations of neurons in such a way that they define 'synchronous.' Spikes from two different neurons are synchronized if they occur within a given oscillatory cycle. (We know this because follower neurons typically spike only at the peak of the cycle.)"

Temporal coding is not a one-neuron affair but coordinated activity across neural populations. Stopfer noted: "If those spikes somewhat had more jitter across projection neurons, you might not be able to get the Kenyon cell to spike." (Kenyon cells are neurons in the mushroom body of insects, a neural structure comparable to the piriform in mammals.) Now, you might not get these cells to spike "because that activity would be distributed and wouldn't add up to meet the threshold needed to make the Kenyon cell spike. The synchronization is allowing the Kenyon cell to spike when enough projection neuron spikes are lined up in time. But remember, there's also this very slow pattern of inhibition that's determining which projection neurons are firing at what time relative to each other. Only a certain population of projection neurons is going to fire even at roughly the same time because some are going to be inhibited, and some are being excited."

Spiking patterns in neural populations build on each other, resulting in increasingly complex interactions from signal propagation to motor behavior—and back again. Spiking resonates with behavior in action-perception coupling. How does the brain read these signals?

Interpreting the Morse Code

The essential stage show has three acts. First, specificity in neural responses increases with stimulus exposure. Stopfer detailed: "The first time you present an odor, there are no oscillations. You have to present an odor two or three times before the oscillation starts to build up. That's because there's this activity-dependent plasticity that takes place within the antennae lobe. The local neurons that are activated become more effective over repeated activations. The inhibitory local neurons become more and more effective at synchronizing the projecting neurons over the course of repeated odor presentations. We think that's enabling the system to become more specific as the odor remains present."

Second, this leads to selectivity and categorization with repeated experience. Stopfer added: "What's important and what's noise? Maybe one way to decide what's important is if it sticks around, if it's not very transient. If you're close to an odor source, or if it's something that your receptors are especially attuned to, then encounters with that odor will be repeated. That will drive this process to build up, allowing the rest of the system to become more specific. The way we think this plays out over time is the first time you encounter an odor, you give off this big burst of activity that's not very specific, and maybe that gives rise to something like an orienting response signaling there is something new there. If the odor continues to be around for a while, the system is driven to specialize in that odor. In terms of dynamical systems, you could say the antennae lobe is forming an attractor around that stimulus [marking a state space], and then the system is able to specify what that odor is more precisely."

Third, repeated experience facilitates the refinement of sensory coding. Stopfer continued: "You go from a very general response to a much more specific response. At the very beginning, you get this big burst that tells you there's something novel in the environment. Then right after that, you start to categorize it: it smells floral versus savory, for example. If the odor is still present, the system becomes more and more specific as this process builds up, the response downstream becomes more specific, and then you can identify exactly what it is. The same circuit at first will give you this generalization: it's something [fruity]—and then that same circuit over time will say: oh well, it's cherry, not strawberry. It only happens if the odor is there long enough to perhaps be of interest to the organism."

Notably, these mechanisms structure and prioritize input bottom-up. "That's an example of plasticity that entirely feeds forward," Stopfer argued. "It doesn't rely on an attention mechanism coming down from other parts of the brain" although top-down effects certainly guide, fasten, and enhance such a process of observational refinement. And so the brain's activity measuring neural responses is shaped by several dynamics, including directionality (feedforward and feedback) and threshold (influenced by the neural experience with specific signals).

These mechanisms, regulating the behavior of neural populations, explain how the olfactory system further differentiates odors that are either ambiguous or underdetermined in their combinatorial coding and signal distribution. Their iterative evaluation facilitates selectivity and learning processes that result in observational refinement. "What this doesn't explain is how meaning gets attached to odors," Stopfer concluded. "To me, that's the biggest open question—because it turns out that the connectivity from the projection neurons to the Kenyon cells is largely random."

Signal distribution seems random. But its measure is far from arbitrary, Stopfer concluded. "In olfaction, the relative timing of different neurons carries information."

General Principles, Individualized Representations

Neural representation in olfaction is individualized. There is no stereotypic, topographical mapping of odor, some generalizable order that links physical stimulus space to perceptual space in the brain of an organism. That does not render smell without rules. The objective measure is the computational processes coding signaling bits and pieces into revisable perceptual judgments—an ongoing response proportional to the physiological conditions of the perceiver and the changing stimulus ratios in the environment.

Olfaction is dynamic in coding and computation. Its neural expression operates by individual, rather than fixed, representation. McGann agreed: "Neural codes develop over time, and the brain internally knows its own code. There's a sense in which I don't necessarily need to find the one code that's true in each mouse or each in person because they're each going to have, potentially, their own individualized representations that are maybe unique to that individual based on previous experience, based on developmental instances, things that happened." And so, next to organismal evolution, development and experience provide the background against which the brain as a measure of the world is calibrated. Brain activity, as a measure of the world, is self-organized and selective.

Perceptual representation in such a processual framework is about informational content. This content does not necessarily represent perceptual instances as classes of universal "perceptual objects." The olfactory brain measures "odor situations" to evaluate how cues are related to each other (temporally, combinatorially, causally), and to attribute these perceptions a specific value (pleasant, putrid) and behavioral response. The informational content manifested in these odor situations is variable. It depends on associations formed between input ratios and combinations, next to the appraisal of their (anticipated) interactions.

This changes the ontology of the system entirely. And it changes its analysis, Kay emphasized: "You have to be comfortable with the fact that you're not studying the same system every moment. You're studying a dynamic system. You're catching it at some point, and you have to understand where your snapshot is." This yields experimental challenges—and a window of opportunity.

This deep intertwinement of input with output processing in olfaction creates a potent interface to bring wet-lab neuroscience together with computational theories. Afferent connectivity (in the piriform) and temporal coding (primacy and population coding) present cellular expressions of the two main principles in recent theories of the "predictive brain." These theories, as we heard earlier, assume that your brain learns stimulus regularities. It measures the environment via prediction, where perceptual content matches current input with prior experience. Its high plasticity and contextual encoding make olfaction an excellent model to analyze how the two computational principles of anticipation (top-down) and error correction (bottom-up) link to cellular mechanisms at the periphery and in the brain.

Psychophysical studies long suggested that olfaction operates by predictive principles. Jonas Oloffson highlighted cross-modal effects: "When you get the cue 'lemon,' and your job is to say whether the following odor is lemon or not, you are faster when it is actually lemon than when it's not lemon. We theorize that the reason for that is that we construct a predicted [temporary] template in the olfactory cortex that then matches. That facilitates the perception of that quality, that specific odor object. When there is a nonlemon odor, the brain has to figure out what kind of object it is. If it's primed by a label, the response is facilitated in terms of reaction times. In that scenario, olfaction is not about the perceptual change, but validating the nonolfactory cue."

Recent neuroimaging experiments in the piriform and neighboring regions support this interpretation. When Jay Gottfried picked up on the idea of the predictive nose, he put it to the test. In 2011, Gottfried and his postdoctoral researcher Christina Zelano demonstrated how predictive processes direct olfaction in humans.[22] They compared activation patterns with functional magnetic resonance imaging. Gottfried and Zelano first familiarized subjects with a distinct smell (a nasal mask ensured controlled administration). Participants were asked to identify this particular odor in subsequent exposures to three kinds of stimuli. Throughout these trials, the given stimulus sometimes matched the anticipated odor quality, sometimes it was another odor, and sometimes it was a mixture of both. (The study involved two groups paying attention to one of these two odors.) When the stimulus matched the expectation, pre- and postexposure activity showed higher correlations compared to trials in which the subjects were presented with an unexpected odor.

These results indicate that the piriform creates temporary anticipatory odor templates, allowing for quicker identification of subsequent familiar smells. Additionally, temporal differences in some cortical domains (such as the anterior piriform cortex and orbitofrontal cortex) showed anticipation patterns that persisted for several seconds after stimulus exposure, regardless of stimulus nature. This suggested that activity in these domains was instructed primarily by expectation rather than the given stimulus. Meanwhile, other domains (such as the posterior piriform cortex) displayed faster transitions from pre- to poststimulus patterns, signifying a functional role in the integration of feedforward and feedback connections. Here, processing was dominated by input rather than feedback connections. Further exploration is ongoing.

At present, it sounds apt to start rethinking perceptual effects as a measure of variable stimulus ratios, with their meaning informed by selective biases introduced by experience and behavioral responses. The olfactory system is strikingly flexible in its coding responses and perceptual expression. Stimulus evaluation is sensitive to minute differences in exposure and experience, affording the integration of olfactory cues into multiple decision-making contexts. This also shows that variations in perceptions of the same physical stimulus are not a failure at the level of sensory detection, but a constitutive part of the processing system.

A Psychoneural Theory

Perceptual space in olfaction is computed as experiential space. The sense of smell operates as a dynamic measure to evaluate information in context, namely as changeable mixtures. Flexible permutations of signaling bits and pieces are made possible by the particular setup of the olfactory pathway, starting with distributed receptor combinatorics in the epithelium (Chapter 6), representation of environmental statistics at the bulb (Chapter 7), and individual computational scaling at the neural level (this chapter). This is how the brain measures the world via the nose. Naturally, these coding principles find their expression in the perception of odors.

What might such a processual perspective on olfaction entail in terms of psychology? Chapter 9 will give an answer. It picks up on the central processes governing olfactory coding on a neural level, as analyzed in this chapter, and links them with psychological mechanisms: iteration and observational refinement, learning and memory. The model of the brain as a measuring machine thereby integrates neural principles with perceptual effects; it offers a framework in which molecular science meets perception.

The structure of phenomenological experience, what it's like to perceive the world, is not independent from the neural architecture that creates it. How the mind works is fundamentally an expression of neural processing—while, at the same time, neural processes hinge on our behavioral interactions with the world. Kay concluded: "If you think of it from a dynamical systems perspective, it's like the behavior is the attractor. The behavior coopts the system, and the emergent mind feeds back down. The emergent mental concepts or ideas impinge on the lower-level activity to organize it into something that accomplishes the behavior. Because the evolutionary pressure is not on the neurons, it's on the organism."

CHAPTER 9

Perception as a Skill

Imagine being part of the following experiment. In front of you are two jars, indistinguishable by visual appearance. Each contains an aromatic mixture of identical color and amount. Only their labels are different. One reads "Parmesan," the other "vomit." Do they smell the same or different to you?

Parmesan sounds more agreeable than vomit. The images associated with each label refer to different things in our experience. Parmesan is food; vomit is a contaminant and disease factor. Separate mental images, different qualitative impressions, and opposite hedonic value. Now, what if you are told that these two mixtures are identical, that each flask contains a blend of butyric and valeric acid? Would people identify them as identical by their smell, or get fooled by the labels?

In 2001, Rachel Herz and her student Julia von Clef tested people's reactions to five identical odor pairs, administered with different labels, and measured the influence of verbal labeling on the perception of odors.[1] In addition to the butyric and valeric acid mixture ("vomit" versus "Parmesan cheese"), the odors and respective labels presented were patchouli ("musty basement" versus "incense"), pine oil ("spray disinfectant" versus "Christmas tree"), menthol ("breath mint" versus "chest medicine"), and violet leaf ("fresh cucumber" versus "mildew"). Participants' responses were clear. The majority (83 percent) believed that the odor in each pair was different as a function of its label. Participants were also consistent in assigning opposite hedonic values to each odor pairing according to whether the label given for it was seen as pleasant or unpleasant.

Herz and von Clef presented their results as evidence for something like an olfactory "illusion." The notion of illusion implies various things. One meaning of illusion is deception: the divergence of a perceptual representation from how things "truly" are. Herz's aim was to demonstrate the causal effect of context on the content of perceptual experience.

Context matters to the categorization of odor. "Absolutely, 100 percent," Herz nodded. "You need to have this extremely flexible system, and context is absolutely critical. That's exactly the level at which I talked about the illusion." Naturally, context matters to illusions in other senses, too. Herz highlighted the popular Müller-Lyer illusion (in which two arrows of the same length appear to be of different size, based on whether the arrow's ends are bound inward or outward). "The arrows feathering out or the arrows feathering in are the context. Depending upon the way the arrows feather, the line looks longer or shorter, but the line is actually identical in both cases. The verbal context is like the arrow feathers. I can make the butyric and valeric acid mixture be vomit, or I can make it Parmesan cheese as a function of changing the context. I'm using words to change the context. The Müller-Lyer illusion uses the direction of lines that surround that central line to change the context."

It is intuitive to think that both mixtures should have the same experiential quality since the mixtures are identical in physical features. But what is the "real" perceptual identity of butyric acid: vomity, cheesy, both, or neither? In the end, the smell of butyric acid is no more akin to vomit than it is to Parmesan cheese or vice versa. These molecules are promiscuous regarding their source materials; they are part of vomit as well as Parmesan aroma (next to gingko tree seeds, or noni fruit). So the shared sensory identity of these mixtures resides on a nonconceptual level.

On a conceptual level, these stimuli signify distinct materials with different affordances. Cognitive processes, such as semantic framing, shape the perceptual meaning of physical input. This psychological expression resonates with the autoassociative structure of the olfactory cortex where topographically widely distributed signals were integrated further downstream (Chapter 8). These signals can interact with various parallel processes simultaneously and form perceptual categories via top-down effects from higher cognitive levels, thereby accounting for such divergence in perceptual outcomes.

Many molecules occur in all kinds of circumstances and chemical environments, where they significantly change informational content and value. There is no unique "intrinsic meaning" or "representational concept" encoded with the isolated stimulus to be abstracted from the contextual experience of an odor (Chapters 3 and 4). The human brain learns to associate a stimulus with different semantic concepts, routinely aided by additional cues (verbal, visual, etcetera). The brain, on this account, is able to "interpret" the qualitative disposition of butyric acid as Parmesan or vomit, depending on context. "Absolutely," Christian Margot agreed. "You can change the fragrance by experience and by context." Steven Munger replied: "One of the things that we often don't consider is how we might differentially perceive an odor in the context of experience. The same odor can have different meanings depending on your history with it."

The Stain of Subjectivity

Olfaction shows a higher degree of perceptual variation in comparison with other senses. The same odorant can have various meanings in the perceptions of different people, and even for the same individual. Its considerable variability appears to imply an intrinsic subjectivity of smell and the reason it seems so difficult to arrive at a general cognitive theory of olfaction. The previous chapters revealed a different picture. We saw that the coding principles and distributed representation of the olfactory system support several ways of integrating an olfactory signal (notably, already at the periphery).

Information of the same physical stimulus and its features partakes in alternating yet parallel information processes, shifting its perceptual integration and, in turn, behavioral meaning or conceptualization. It's an essential condition for the sense of smell to do its job: facilitating flexible and fast-enough behavioral decisions in an unpredictable environment. The perceptual content of odor experience depends more on context and exposure conditions than on perspective-invariant object recognition of particular features (as in vision).

Variability is not the same as subjectivity. The latter implies the absence of an objective measure. Variation in olfactory perception, however, has been shown to arise from the mechanisms of odor coding (receptor level) and the computational basis of signal integration (central processing). This constitutes the objective basis for studying variation as a measurable effect of the underlying causal factors. To be sure, some degree of variation also occurs in other senses. The point thus is a general one, namely that (1) the higher degree of variability in olfactory responses mirrors the coding principles of the system, and (2) that differences in causal grounds and effects between the senses must be examined against this backdrop. Perceptual variations in olfaction do not constitute deviations to be "explained away" as subjective distortions. They are the hallmark of its functioning.

Smells, therefore, are polysemic by design. Olfactory stimuli act as signs that can signify many source materials and connotations. Their perceptual content is determined by the context of exposure, but also the demands and constraints of the system. Essential to understanding the inherent perceptual variability in olfaction is that there are genuinely different ways of "reading" the same stimulus and treating it as radically different kinds of information (food versus contaminant, pleasant versus unpleasant). Such promiscuity of the olfactory system prompts the question: How does the mind make sense of scents?

The Cognitive Mapping of Odor

Whether a rose by any other name smells as sweet may not be as evident as we think. Odor perception is inherently ambiguous. And, as we know now, ambiguous means conceptually underdetermined, not inaccurate. Olfactory information, on its own, lends itself to be sufficiently ambiguous concerning its ecological sources, biological coding, and neural representation, as well as mental imagery. Analysis of the psychological mechanisms implicated in the cognitive mapping of olfactory experience ought to reflect that issue.

Odor images are not encoded in the stimulus; they constitute mental impressions that arise from the categorization of sensory information. What processes regulate this categorization? Prior chapters focused on the physical basis of information extraction and distribution in the system. They explained that modeling isolated structure-odor responses is like defining a species purely by its morphology, not its ancestral origins. Now we must ask: How do we still end up with reasonably distinct and robust perceptual categories, like rose, apple, and pee?

Categorization is a process that requires perspective or a point of reference. The ability to evaluate a stimulus in isolation differs profoundly from the evaluation of two (or more) stimuli in relation to each other. In 1956, American psychologist George Armitage Miller introduced this difference as the distinction between relative and absolute discrimination.[2]

Absolute discrimination is the ability to identify a stimulus in isolation—for example, when you identify a given odorant as minty and describe its quality as being green, cool, and fresh. In relative discrimination, the task is to tell whether two stimuli are the same or different (or whether one or two stimuli differ from a third). Here, your brain measures the perceived "distance" between stimuli. The smallest distance between two stimuli—before perceived as being identical—is called the "just noticeable difference." These two processes, absolute and relative discrimination, perform different kinds of perceptual judgment, with measurably different functions and performance.

People are extremely good at relative discrimination in smell. Trained or untrained, the human ability to tell different odorants apart is strikingly accurate. Our nose is tuned to most minute differences, molecular and perceptual. Psychophysical studies, testing odor discrimination, demonstrate that the human nose responds to elements of incredibly low quantities and dissimilarities. Just how good are we? A mind-blowing example is TCA, trichloroanisole ($C_7H_5Cl_3O$), the component responsible for the aroma of corked wine.[3] Some people have been measured to detect TCA in the parts per trillion! We're fantastic at smelling.

On the level of molecular detection, sans conceptual labels and identities, the human nose is a powerhouse. "Let me tell you how sensitive your nose is," Terry Acree noted. "When we manufacture a standard solution of an odorant, all we have to do is to make another one, and it will smell different every time. We cannot make two odorant solutions to smell exactly the same."

Things get messier with absolute discrimination. Identifying a smell without cues is notoriously hard for the untrained nose. Out of context, it is challenging to describe or name an odor—even when you are familiar with it. Moreover, people differ widely in their semantic associations with decontextualized olfactory encounters. These differences in perceptual evaluation are an expression of the cognitive conditions under which we assign odors conceptual content.

A story illustrates this point. The psychologist Trygg Engen described an incident from the 1980s.[4] "A large company with a bestselling brand of fruit-flavored drinks by mistake once labeled the strawberry drinks as raspberry and vice versa." The company feared its mistake would result in a barrage of consumer complaints. The complaints never came in. No one even noticed the error. "Although these flavors are different when compared side by side, they are easily confused when encountered singly." This incidence is more than anecdotal. Bill Cain and Bonnie Potts documented a similar phenomenon in 1996. They examined whether participants would realize that the odorants in their test were secretly replaced with other odorants.[5] For example, they exchanged garlic with vinegar, orange with lime, and soy sauce with molasses. The panelists took the bait! Participants failed to recognize that the test odors had been replaced.

Such results sound puzzling compared to our high accuracy in relative discrimination. "This comes back to my experience in the applied world," Avery Gilbert confirmed. "I've asked plenty of people to describe perfume formula, perfume fragrances, or this fragrance, or that fragrance. People who suck at describing it immediately know what they like and don't like to a decimal point. They'll give us a 'this an 8.5,' 'this is like a 6.3.' And they're very *consistent* about it! They can make very fine discriminations like, 'This one's a little spicier.' They can tell when it's different. So discrimination is effortless and very good."

Functionally, relative and absolute discrimination are apples and pears. Relative discrimination detects and measures sensory change through difference; absolute discrimination involves the assignment of conceptual content, such as the quality of an object, an object type, or a scene. The former process is a sensory evaluation of stimuli features in comparison with each other; the latter involves a cognitive embedding of the stimulus in relation to prior encounters. Absolute discrimination includes memory.

One of the notable things about our generally low abilities for absolute discrimination in olfaction is the fact that the opposite is also possible. Just because many people struggle with absolute discrimination does not mean it designates a cognitive inability in human olfaction. Odor experts excel at this task: consider perfumers, fragrance chemists, or wine and whiskey tasters. Their skill is acquired through explicit and iterative conditioning, built on refined observation and categorization. For the study of cognitive capacity in olfactory categorization, and how it mirrors its physiological basis, we should look to the experts.

The Secret of Expert Noses

It is not uncommon to use skilled participants for experiments. The choice between experts or novices depends on the aim and procedure of the experiment. What is it that you want to study? Are you exploring a statistically significant behavior in everyday performance, the details of a specific mechanism, a proof of principle, or something else? Some experimental designs present more reliable and stronger data because their data are gathered from expert participants. An excellent example is a 1942 experiment by Selig Hecht, Simon Schlaer, and Maurice Henry Pirenne. They discovered that retinal receptors detect single photons (although the retina requires a higher threshold for the stimulus to elicit a conscious effect).[6] Remarkably, this insight first came from a psychophysical experiment. Its results were confirmed physiologically almost forty years later![7] The test itself was considerably tricky and required an extensive focus on the task by the subject (the experimenters were taking turns)—a strenuous exercise that might not have succeeded with naïve and less involved participants, such as the average college undergraduate.[8] That is why we also should look at experts, especially in olfaction.

The first thing to recognize is that these experts do not have super noses that detect stuff beyond what is ordinarily available to humans (like Jean-Baptiste Grenouille in Patrick Süsskind's novel *The Perfume*). Scent experts do not have hidden genetic features, enabling them to smell something inaccessible to the average nose. It is not a matter of biology.

The winemaker Alison Tauziet confirmed: "It's interesting to me that when I'm tasting wine with people, they assume I know way more about olfaction than they do. People underestimate their skills." For Tauziet, it comes down to a difference in the training of relative and absolute discrimination: "If you put two different things in front of people, they say: 'These are different things.' But the development of the language attaching it to that smell, that's the only thing that's distinguishing me. That is a skill that needs developing. I think that everybody has that capacity if they try."

Experts engage in active learning, meaning the targeted refinement of observation through the cognitive engagement with odor. Christophe Laudamiel added a perfumer's perspective: "We don't have a super nose but we notice things; it's our job to recognize a lot of smells. We pay attention. We recognize what we smell. We know how to describe things, and we know how to compose."

This ability takes training. It takes attention, the cognitive development of perceptual skill, and time. It demands long-term attention to the cognitive structuring of perception, next to sufficient knowledge of the material origins of odor.

Harry Fremont at Firmenich looked back at his own experience and laughed: "I told you I needed twelve years to be comfortable in my job! And part of it was this mastering of those ingredients. Not only from the perceptual and language point of view but also concerning an understanding of the effect in the formula point of view. It's the experience, I think."

The study of expertise reveals something important about the brain: the olfactory brain is highly plastic. It provides a great model to examine the effects of individual experience and general training on neural processing. Recent studies have shown that olfactory training leads to significant structural modifications in the brain.[9] Johannes Frasnelli explained the scientific interest: "We did some studies where we saw that there is a link between the structure of certain brain regions, like the thickness of the cortex and the thickness of the grey matter layer in certain brain olfactory processing regions, and the ability to perceive." Looking at a number of test scores, Frasnelli's team found that people with better perceptual capacities had a thicker cortex. They also looked at "people who have lost their sense of smell. And we see that they have a reduction in these centers." And what about people born without a sense of smell? Do they show a difference in cortical features? "So, we did all of these studies where we compared different groups to each other."[10] All groups showed differences in the brain. But how innate are the abilities of experts? "Is this something because they're born this way? Or it's the training causing the brain to change itself?" Not much later, in 2019, Frasnelli's team found that undergoing just six weeks of olfactory training results in changes in cortical thickness "in the right inferior frontal gyrus, the bilateral fusiform gyrus, and the right entorhinal cortex."[11]

Olfaction is a sense with an astonishingly high disposition to expand its capacities through targeted experience. The unparalleled abilities of expert noses build on general processes of perceptual learning. Humans can develop a powerful analytic faculty of odor perception and evaluation. Olfactory experts differ from untrained noses because they display a remarkable technique of structured cognitive engagement with sensory stimuli that drives their sophisticated perceptual skills.[12] What separates perception from cognition is indeed not more than a feeble, permeable line drawn and redrawn by repeated use and practice.

Perceptual Expertise as Observational Refinement

Experts pay attention. Specifically, experts know how to pay *selective* attention. Not only do they know what to look for, they also have an advanced catalog of potential options, criteria, and concepts on which to base their perceptual judgment.

Watch and listen to perfumers or sommeliers do their job. A scene in the movie *Somm: Into the Bottle* perfectly encapsulates that experience.[13] This 2015 documentary follows four sommeliers on their journey to passing the challenging master sommelier exam. One scene is paradigmatic: We see the sommelier Ian Cauble evaluating a wine, uttering qualitative descriptors in a rapid, incomprehensibly fast sequence.[14] Cauble, looking at a glass of white wine, begins:

Wine one is a white wine, clear star bright. There's no evidence of gas or flocculation.

[Cauble holds the glass at a 45-degree angle, analyzing how the wine breaks the light at its edges.]

The wine has a light straw core, consistent to green reflections in the edge. Medium concentration of color.

[He pauses briefly, then swirls the wine in the glass, sniffs, and continues describing its orthonasal character.]

We're almost coming out of like this lime candy, lime zest. Crushed apples.

[Pause, swirl, another sniff.]
Underripe green mango. Underripe melon. Melon skin.
[Sniff.]
Green pineapple.
Palate.
[Swirl, then he takes a sip, forcefully rinsing the wine in his mouth, spits.]
The wine is bone-dry. Really this, like, crushed slate, like crushed chalky note, like crushed hillside.
[Looks at the wine from a 45-degree angle again.]
There's white florals, almost like a fresh cut flower, white flowers, white lilies, no evidence of oak.
[Swirls the wine, takes another sniff.]
There's a kind of that freshly opened can of tennis balls, and like a fresh new rubber hose I get.
[Pause.]
Structure.
[Next sip, next spit.]
Acid is medium-plus. Alcohol is medium. Complexity is medium plus.
[Continues to swirl the wine in the glass, which is now on the table; looks at the color.]
In this conclusion, this wine is from the New World, from a temperate climate. Possible grapes are Riesling. Possible countries are Australia. Age range is one to three years. I think this can only be one thing.
[Final sniff.]
This wine is from Australia. This wine is from South Australia. This wine is from Clare Valley, 2009 Riesling, high-quality producer.

Cauble arrived at his verdict in about a minute. He was correct. To many viewers, this scene gives a wonderful glimpse into the obscure subculture of sommeliers—a world that does not bear much resemblance to the everyday consumption of wine. (And this is not just because the wine was spit into a bucket.) The almost impenetrable, idiosyncratic language ("kind of a freshly opened can of tennis balls, and a fresh new rubber hose") makes one feel like a mere outsider looking in.

Nothing in this scene is about words. That is critical to understanding Cauble's perceptual abilities. Forget the tennis balls, forget the rubber hose. These particular labels are not explanatory of the wine's qualities but act as cognitive handles in Cauble's deductive tasting. When Cauble says "no evidence of oak," it is a cognitive pointer. Acree clarified: "The wine was not stored in oak, which is common for Riesling made in Germany." This wine, "it's made in Australia, in stainless steel. You don't have an oak aroma in Australian Riesling. And then, here is guaiacol, which is one of the major components of . . ." He laughed, "it's not rubber hose, but it's plastic hose. It's Band-Aid aroma, and also in things like tennis balls: a polymer-based product. You open a can of tennis balls, you get this smell, and Riesling can or cannot, depending on how it's made, end up with guaiacol."

Look at what Cauble does. The real work is the cognitive sampling in his deductive processing, how Cauble structures the series of perceptual impressions during testing. Sommeliers, like Cauble, are trained to use a cognitive grid, containing an encyclopedia of wines and wine properties. Each wine, region, and year is memorized in its specific characteristics. It is like learning the grammar, vocabulary, and semantics of a new "cognitive language." This grid, with its vocabulary, acts as a memorization framework to structure perception.

If you consider the scene in *Somm* again, you now can observe several strategic operations at play. Wine tasters begin by holding the wine glass at a 45-degree angle to assess what is called the body of the wine. If the wine is red and, say, has a lighter magenta color, it might be a pinot noir. Or, if a white wine appears yellow in its undertones, it could be a chardonnay. How the light breaks at the surface of the wine, whether the surface contains a slight oily film, and other visible markers already provide a list of properties that allow the tasters to form an initial, rough hypothesis about the kind of wine at hand. That is also why wine lovers swirl the wine. Swirling increases the liquid's surface and releases more aroma molecules, revealing more of the wine's bouquet. Try it. Pour yourself a glass of wine. The effect is especially strong in red wine. First smell the wine unswirled. Then swirl and smell the wine again. There is "much more" aroma the second time! This is not magic, merely surface physics and chemistry.

Cauble's list of aromatic qualities was not open-ended or extensive; it was confined to three basic attributes: lime candy, lime zest, crushed apples. This limited set of aromatic notes resonates with the Laing limit (Chapter 8). Acree, a big fan of Laing's work, emphasized the significance of the Laing limit for us again. Laing originally "used individual components to test whether or not people could identify these individual components in the mixture. I think they were given one to five chemicals out of seven . . . and they were trained to associate a word with each of those seven chemicals." But that was not the key to understanding the Laing limit.

"Laing then reproduced the entire experiment in which he trained people to recognize mixtures of chemicals as natural products. Like lily extract, or bergamot extract, or *Rosa damascena* extract. These were words they associated with collections of chemicals. Then he mixed them together and got the same result! They could only detect two or three of these mixtures within the mixtures. Think about what that means. Now we're not talking about individual chemicals, we're talking about patterns of chemicals." Acree emphasized that when "mixed together, the components were creating hugely different patterns—still, sommeliers were able to recognize *this* pattern and *that* pattern . . . but only two or three of them. That's why I think: let's not talk about image formation but image recognition."

In the movie, Cauble foraged for recognizable patterns matching his memory of wine aroma profiles. The three categories picked by Cauble provided him with a preliminary range of wine profiles, which he used to hone in, attending to some of the wine's more specific qualities. Then he followed a decision tree of neighboring qualitative notes to sample and explore the wine's profile: underripe green mango, underripe melon, melon skin. It is a method of iterative attention and specification. Cauble arrived and fixated on the odor note "green pineapple" (in relation to the hypotheses forming in his mind). Acree explained: "There is apples, mangoes, and he says crushed apples and underripe mangoes, all right? Melon skin and pineapple. All of these things like apple, mango, melon, and pineapple have present in their volatile mix high levels of ethyl hexanoate. In combination with other components, ethyl hexanoate can create an image of any of these three to four things.

"If you have a collection of key odorants, say six or nine key odorants . . . certain collections can create what is called a 'clique.' It's a subset of the total picture. And the subset has its own character. If we have a wine that has nine key odorants that make up its odor spectrum and you sniff it, your brain can take three of those and come up with an odor image for those three. It could take another three components overlapping, and come up with a different odor image. Or another three and experience a different odor image.

"That's exactly what is going on when the sommeliers . . . when these guys are sniffing the wine. They smell lilies, they smell roses, they smell tennis balls." Acree drew Figure 9.1 to illustrate this point. "What they're doing is they're taking subsets of odors as a clique and seeing these subsets as representing a different image. That's why you can have these overlapping images coming out of a single mixture. The process of deductive sniffing is not finding

Sommelier—Riesling

Aroma (orthonasal)

Lime (candy)
Lime (zest)
Apples (crushed)
Mangoes (underripe green)
Melon skin (underripe)
Pineapple (green)

Aroma (retronasal)
[bone dry taste]

Slate (crushed)
Chalk (crushed)
Hillside (crushed)
Flowers (fresh cut, white)
Lilies (white)
    Oak (none)
Tennis balls (fresh open can)
Rubber hose (new)

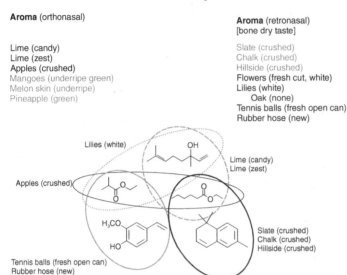

Lilies (white)

Apples (crushed)

Lime (candy)
Lime (zest)

Tennis balls (fresh open can)
Rubber hose (new)

Slate (crushed)
Chalk (crushed)
Hillside (crushed)

Figure 9-1 Perceptual cliques and overlapping chemical groups. Illustration of the aroma patterns in Riesling as described in *Somm: Into the Bottle,* matching perceptual to chemical profiles (explanations in text). Source: © Terry Acree, reprinted with permission.

chemical components, it's finding perceptual components, or perceptual views. Essentially, it's like selecting words from a sentence . . . to make different sentences. It's a Scrabble game in which you not only get to make words from letters but compound words like *das Fingerspitzengefühl.*" He laughed. "Maybe smell *was* designed by a German chemist!"

Wine tasting is deductive perception. Following orthonasal smelling, Cauble double-checked his emerging hypothesis about the type of Riesling with other modalities. Switching from olfactory qualities to another modality, describing mouthfeel and acidity, Cauble cross-examined his gradually refined assumption about the wine's identity. By now, Cauble has arrived at possible options and started to exclude items, testing and filtering his observations about olfactory notes in the wine (at this point, you see Cauble swirling and smelling again). So the film scene depicts a complex, cross-modal procedure of decision-making. Cauble iteratively refined his observations over several cognitive steps through targeted perceptual engagement with the stimulus (the wine).

Cauble's is just one method. To analyze the nature of expertise, the decisive point of focus is not the specific structure of the grid, which varies between wine schools. Effectively, the whole process is about the structuring of perception through an acquired cognitive system. Tauziet explained: "I smell it and taste it on the whole, and then . . . narrow the options. If you have something that's super ripe, like some riper fruits or dark black fruits, then you know you are in a warmer area. If you taste it, for example, if you have a lot of acid versus low acid . . . that tells you also about warmth. That starts to nail down what parts of the world it could be grown in. Freshness or evolution starts to tell you about time spectrum. And then . . . you can kind of hone in on probably this vintage because this kind of weather happened, or that vintage because that kind of weather happened. So I have a few ways of looking at the wine to put me where I am in the world. I go towards 'where' and 'when in time' it is."

Skilled olfaction tells that perception comes in layers. The sensory complexity of olfactory stimuli is not accessible immediately. It requires access through selective and iterative attention. Perceptual expertise fundamentally involves refined judgment through targeted observation. It is a skill, which builds on multiple evaluative dimensions shaping the content and structure of an expert's olfactory experience. Two features of olfactory learning are particularly relevant. Guiding observational refinement and underpinning perceptual content formation in trained and ordinary perception are language use and domain specificity.

Language as a Cognitive Handle

Experts acquire language to frame observations by connecting their perceptions to a memory grid of prior experience. It is a tool or cognitive handle to structure, fix, guide, and shift attention. That means it is not the choice of words or descriptors that matters, but their use, or the speech *act*.

Odor language is inherently communal. Expert language, like all language, aims to communicate categorical or abstract content next to subjective sensations. It is an ongoing negotiation in which contextual odor experience is integrated into a broader cognitive landscape, not only of the individual but of a community. Laudamiel remarked: "Among perfumers, you learn the vocabulary already during your training. We do introduce new words frequently, not every day but frequently, because we have new smells all the time that we have to describe. We love that someone sees what we see or what we smell. If I smelled it in France, I love that the perfumer in New York can see a little bit of what I've seen."

Communication, as a speech act, is based on agreement and requires a sufficient level of convention. The specifics of descriptions may vary while particular standards of reference exist. These standards depend on the level of learning and expertise (Are you learning the basics? Or have you had some chemistry training?), the context of an application (To whom are you talking? What is the purpose of communication?), next to other pragmatic factors. Say you want to start learning about how to identify and describe wine aroma. You can buy wine-testing kits with which you can train yourself to recognize about fifty odors typically used in red or white wine, learning to associate their smell with a label via classical conditioning. You may also acquire an initial olfactory repertoire with so-called fragrance or aroma wheels. These wheels showcase a set of primary odor qualities—arranged in a circle—for highly specific domains: systematizing odor notes in red wine, white wine, whiskey, brandy, beer, and *Cannabis*. In fact, there is a fragrance wheel for almost anything these days—even sewage water.[15]

The founding figure of the wine aroma wheel is Ann C. Noble, at UC Davis, in 1984.[16] Noble shared how the idea emerged: "The problem was that we don't have a vocabulary because, for most people, it's not natural to be thinking about the world by describing its aromas." Noble wanted to give others the tools to access what they could not describe cognitively. "I was teaching at the time," she continued, "working specifically with a technique called descriptive analysis for describing wine aroma, where you have to describe the words that describe the quality. And you can do it with the flavor, with the texture, with the taste, the mouthfeel, and so forth. You find the specific words that describe the attributes. For wine, it's primarily smell, and taste, and mouthfeel. Finding the words, I was teaching classes, teaching students how to do this, and I, of course, am learning on the way. I would always include physical standards so that people could smell it and say: 'Oh, yeah. You know what? That's clove, and that's cinnamon.' Or: 'That's pineapple, and that's bell pepper, and that's what I smell right here in the wine.' It's a rapid learning operation when you link those descriptions to those aromas." Noble's wheel broke ground; it provided more than descriptors. It gave labels with reference standards you could buy at any supermarket (blackberry, rosemary, clove, etcetera). Plus, Noble's wheel came when winemaking in California started taking off.

Perfumery likewise uses fragrance wheels. An example is Michael Edwards's variant; it has been modified several times since first being proposed in 1983.[17] Edwards's wheel includes a set of general qualities: "floral," "Oriental," "woody," "fresh," and "fougere" (derived from the French word for "fern"—featuring smells like lavender, vetiver, geranium, coumarin, bergamot, and oakmoss). These categories encompass subgroups, such as "dry woods" and "mossy woods." U. Harder (working for Haarmann und Reimer) and Givaudan, like others, developed other perfumery wheels.[18] For a professional perfumer, these wheels are too basic, of course, but they aid retail communication. After all, the fragrance industry employs several kinds of specialists in expert niches. Next to perfumers, big companies such as Firmenich and Givaudan employ chemists, retailers, marketing people, and many others involved in the creation, launch, and distribution of perfume (and other scented products).[19] These domains build on various kinds of knowledge, aims, and vocabularies, depending on purpose and background. Different areas need to communicate successfully to arrive at a final product.

Matt Rogers, part of the neurobiological research sector at Firmenich, laid out the ground structure of these big enterprises: "Great question because it's a very fragmented process, and I suspect it is anywhere in the industry. Imagine the final product would be a fragrance for shampoo, for example. There's a lot that goes into that. So, there could be an upstream R&D to discover ingredients that help differentiate that fragrance. They could also develop technologies that help differentiate that fragrance—for example, give it more bloom, a quick onset of powerful odor when you open the bottle. There's technology behind that. That's the upstream part. But then, if it doesn't involve researching new things, you get to the technical development teams—where they're briefed by clients who say: 'I would like perfume that smells like this and performs like this.' And within the technical development teams, there can be perfumers. But there are these fragrance development people, FDMs, that are not classically trained perfumers. They are asked to develop a fragrance that ultimately needs to get validated by a perfumer or a few perfumers."

Fremont, also at Firmenich, detailed the perfumer's view. When talking to people in product evaluation, "they will get more the big picture of what they are smelling, and it will be more about describing the impression: 'Oh, I love the fragrance, it's very fresh, but maybe it's not creamy enough.' Some people at the evaluation use names of ingredients, but very often the good ones are using more general terms that the perfumer will interpret: 'It's powdery, or it's not creamy enough, or it's too sweet, or it's too heavy, or it's too diluted, or it's too green, or it's not fresh enough, or too citrus ... this and that." Fremont leaned back. "The perfumer makes sense of that. Because we don't create in a vacuum. First, we have a request from the client. What we do as perfumers is that we're really translating lifestyle, emotion, colors, anything. Very often we get the requests from the clients, which is for a specific brand. So this brand has some attributes that we know, normally, and a kind of DNA, if you want, of the brand that you need to integrate into your work. The request will be to create, for example, this sporty fragrance, modern, and so on. They will feed us with different information ... boards, photo picture boards, or videos for design. Maybe they show us the last collection. As perfumers, we try to interpret that into an idea."

In the end, olfactory expert language is like all language: learned. It starts with some basics before becoming functionally specialized and refined. Its use adheres to the general conventions of speech. So how do experts make use of language in olfactory learning and practice?

Building a Perceptual Repertoire

The basics of communication are sometimes summarized as a triangular relation between a signifier (word or sign; semiotics), a referent (materials tagged by the signifier), and their mental representation (conceptualization; semantics). This triangle, conceived by the French linguist Ferdinand de Saussure in the early twentieth century, presents a simplified understanding of elements involved in speech acts, excluding pragmatics and context.[20] Saussure's triangle seems intuitive, at least for visual concepts. For example, in color vision you have a word like "red" / "*rouge*" / "*rot*" associated with a referent (a definitive range of wavelengths within the electromagnetic spectrum, here ~700–635 nanometers) and a mental representation (red as a perceptual category). Color vision and language use are more complicated than that, yet this simple schema exemplifies the basic idea.

Odor language is less straightforward, the neuroscientist Catherine Rouby and the linguist Danièle Dubois have argued.[21] Consider the smell example of "rosy." What is its proper referent? The olfactory stimulus, both distal and proximate, is not as discrete and uniform as the visual or auditory stimulus. There is no designated "rose molecule" or rosy feature. Looking for something such as "the" rose stimulus, the perfumer Jean-Claude Ellena wrote, seems an

> innocent approach when we know that the scent of a rose comprises hundreds of different molecules and that none of them smells like a rose. So far I have not found "the" rose molecule, but I have discovered that the smells of flowers have a biologically dictated cycle, and that their composition can vary significantly without them losing their identity.[22]

As a cognitive handle, language allows us to engage with perception through attention. It is a tool for the mind to structure, focus, and revise its access to what is perceived. That happens during olfactory training, where this process shapes perceptual experience. What is perceived is contingent upon experience and learning. Attention is fundamentally instructive in this context. And language provides a matrix by which it is possible to steady "the mind's gaze" on particular features, as selected benchmark referents, for the recognition of feature patterns.

Real olfactory expertise only begins at this point. Knowing names for benchmark odorants does not explain the knowledge and perceptual range of a perfumer. An essential requirement for perfumers is to systematically train and expand their understanding of the way materials give rise to perceptual sensations. This includes chemical behavior, blending patterns, and connections. Fremont nodded: "It's a lot of work, but it doesn't mean that if you memorize all the ingredients, you're also going to be a good perfumer. It's more important to know what the effect of those ingredients is going to be in the formula, honestly." Memorization of benchmark odorants thus "mentally fixes" many common ingredients required to train pattern recognition, such as how materials interact, in what combinations they occur, and what effects are achieved. Laudamiel agreed: "We are very good at pattern recognition."

Pattern recognition involves several layers. There is the material repertoire, a learned catalog of material interactions. Particular ingredients are known to have specific effects in a mixture, effects that cannot be explained by merely adding the features of these single elements in isolation. The whole, in perfumery, is more than the sum of its parts. Laudamiel offered an example. "What about black olive? When you want black olive, it's burnt rubber with wood." Such acquired knowledge of actively blending materials, and building a repertoire of simple formulas, or ingredient combinations, feeds back into the analysis of a finished composition. Fremont gave an example. "The other day, someone came here with something called 'Black Citrus.' It's a shower gel from someone. And I said: 'But that's "Light Blue" with *this* note.' In fact, they analyzed it, and that was what it was. I recognized the combination of the ingredients."

Next to knowledge of a material repertoire is trained association. Association means to create relationships—linking materials with each other, materials with concepts, and concepts with other concepts, while also considering the perceptual background of people in the process. Laudamiel noted that, sometimes, "it's seeing patterns between things that are not related, whether it's ingredients or subjects or concepts or pictures or objects. It seems unrelated, and you have never seen the relationship before. But, because you've learned a certain way, you see new patterns with new objects. We are very good at that." These perceptual associations are part of a learning trajectory, combining a highly specialized, conceptual-material repertoire with active individual engagement.

Successful training and communication rely on the recognition of the different levels by which we refer to (1) various materials, visible and invisible; (2) our own experience, past and present; and (3) the acquisition of competence to connect the previous two. This brings us back to Saussure's triangle, requiring revision (Figure 9.2).

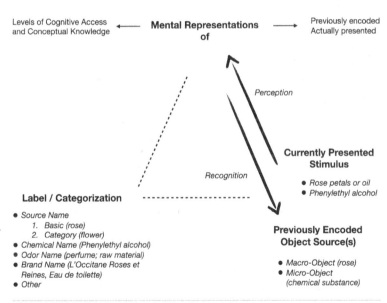

Figure 9-2 Linguistic dimensions of odor. Ferdinand de Saussure's triangle of reference illustrating the relation between a concept (odor quality), referent (odor object), and signifier or sign vehicle (odor name). Source: Reformatted from D. Dubois and C. Rouby, "Names and Categories for Odors: The Veridical Label," in *Olfaction, Taste, and Cognition,* ed. C. Rouby et al. (Cambridge: Cambridge University Press, 2002), 47–66, fig. 4.2, https://doi.org/10.1017/CBO9780511546389.009.

Using language as a cognitive handle to build expertise means coupling different levels of cognitive engagement. First, on the level of conceptual knowledge (representation), there is the temporal gap between previously encoded and currently present experience, which builds on memory and a conceptual frame by which to compare separate experiences as being the same, similar enough, or related by other forms of familiarity. Second, on the material level (referent), it is essential to separate a given stimulus from a previously encountered source object—because stimuli could be either macro-objects or vials with odorants. In the case of the latter, bottles could contain different odorants affiliated with the same mental category—say, octanal (aldehyde C-8) or octyl acetate ($C_{10}H_{20}O_2$) for "orange." Last, there are varying levels of sophistication when it comes to labeling (signifier). A perfumer will know both the general label and category (rose, flower), as well as the chemicals (for example, phenylethyl alcohol) and perhaps even common products in which they occur (for example, as representative of a perceptual signature or pattern).

Connecting these levels is an iterative process in which, one by one, further terms, materials, and representations are added into a growing cognitive landscape. It is like learning a new language; however, odor language acquisition here also entails guided observation, framing your perception.

The learning of olfactory language builds on two general kinds of cognitive processes: knowledge acquisition of how a name relates to a material referent, and the recognition of perceptual content with a mental concept or category. Knowledge acquisition is a linguistic capacity that does not necessitate perceptual representation, just lexical grasp. For example, a perfumer knows that the term "nerol" denotes a monoterpene with the chemical formula $C_{10}H_{18}O$. Now you know that too, even without ever having smelled nerol. Then there is the recognition of perceptual content with a mental concept or category—for example, when you are being given nerol to smell ("Oh, so *that* is nerol!"). These two processes are linked, of course, reinforcing each other iteratively. They are linked first via the psycholinguistic capacity of giving a name to a sensation and learning its proper usage in a specific context, such as perfumery (the scent term T denotes an odorant O, usually associated with a visible material source S). And, to cultivate this linkage, we build on the psychological process of recognizing that perceptual content has been experienced before, connecting it to a similar-enough instance or category.

Focus on olfactory learning reveals that perception and cognition are not entirely separable modules; they are intertwined in practice. Sensory perception should not be considered to present a "given percept," in which sensory impressions merely provide the data for your mind to interact with afterward. Your mind actively shapes what kind of sensory information you grasp, and your experience instructs your cognitive landscape: it is the lens structuring your observations. (And so perception and cognition are not strictly identical but can only be understood in tandem.)

Olfactory experts seem to detect more when engaging with smells; this surplus encompasses complex qualitative details, distinctions, and associations that come with learning as guided experience. Observational refinement fundamentally instructs the formation of odor content and categorization. That is why perceptual expertise is domain specific. An expert winemaker and a master perfumer smell the world differently.

Perceptual Expertise is Domain Specific

Remarkable about olfaction is the range of expertise, how different its conceptual catalogs are, and how strongly specialization affects perceptual evaluation. Olfactory specialization entails knowledge beyond the nose and its training in isolation. It requires the acquisition of other relevant skills. Winemakers do not engage in the same kind of practice as a perfumer. They are not interested in the same features, and their cognitive engagement with materials differs, in the way a basketball player would not automatically be good at baseball only because both sports involve physical exercise and throwing a ball. Jonas Olofsson agreed: "Once you take experts out of their comfort zone, smells that they're not trained for, obviously you'll have a drop in their performance. They will perform more like normal people with a good sense of smell. A lot of expertise might be dependent on the particular format of training, the particularities of their vocabulary, and the stimulus spaces they encounter."

"Not All Flavor Expertise Is Equal," a 2016 study by Ilja Croijmans and Asifa Majid, studied perceptual differences in expert noses. Croijmans and Majid tested three groups: wine experts, coffee experts, and untrained tasters. All three groups were asked to evaluate coffee and wine aromas. The wine specialists outperformed both coffee experts and naïve testers in the identification and naming of wine aromas. (Interestingly, coffee specialists did not exceed in naming and identifying coffee aromas compared to wine experts and naïve testers, although they were more distinctive in their use of descriptors.) People of different olfactory domains only showed a moderate advantage compared to the naïve smeller. Being an expert in one area did not automatically make them succeed in another specialist area. Croijmans and Majid's results suggest that the conceptual understanding of olfactory quality is deeply entrenched in its context, such that olfactory expertise is bound to application domains.

Divergence in skilled performance should not come as a surprise. Different olfactory specialists actively engage with the structure and content of their olfactory perception through the particular demands of their profession: to what their mind pays attention, and how it is trained to pay attention, varies.

Fremont reflected on the divergence of olfactory expertise. As a perfumer, he had always been fascinated by the practice of winemaking. According to him, a central reason for olfactory skill building, and the divergence of perceptual performance, in experts links to their particular interaction with fragrant materials. This interaction concerns handling and degree of control, as well as manipulability of materials. "For wine, you don't know how the season will be. You don't know how wet it's going to be, how cold. You don't know when the grape will be ripe. That's why you have variation; depending on the year the same wine will taste very different. Then I think it's very difficult to manipulate the parameter of the barrel, for example. The condition of the barrel will not necessarily be identical to the previous ones." By contrast, perfumery affords more control and stability in the source material. "In fragrance, we have quality control on every material. You could have variation in some natural product, but we will try to control the variation by mixing different sources. So you even out the quality. Molecules normally should be strictly identical. And then you decide how much you want the fragrance to age because that's very important."

A wine expert knows how to individuate and differentiate individual notes in a complex wine aroma. She knows the aroma profile of wines with which to work. In comparison, a perfumer does not have wine profiles but thinks more in terms of elemental compositions of effects and how to achieve these effects with specific ingredients.

Tauziet gave an example: "When I worked with a young woman that designed a perfume inspired by our wines, I was explaining to her what I smell in the wines: bay leaves, sage, dried sage versus fresh sage, that kind of thing. And she's trying to put that to a compound, to a specific compound because she's building a formula, breaking it down for me."

Winemakers think of wine and its quality in a holistic context. A raisin note might be a sign of a heat wave, indicating the region of a vineyard in a particular year, for example. Perfumers think in formulas, which is an entirely different story. It is hard to make sense of what perfumers observe in a formula as an outsider. Perfume formulas look like lists of ingredients and their proportions. How much of this gets into what, and in what precise amount. These formulas usually are proprietary. Other experts know how to read them. It's like an elaborate code. When you see a list like that, your mind cannot capture what a perfumer is trained to see. It is like saying that a book contains words. Compare it to an algorithm, a musical annotation, an X-ray or brain scan—images that require conceptual knowledge about how to be read.

Hidden in these formulas is an understanding of how to link formulas to their effects. It helps perfumers to record and characterize the subtle relationship between materials and their fickle perceptual qualities. The answer, as usual, eludes a straightforward mapping, also because formulas contain unknown causes. "When you have a formula, the formula often is complex," Laudamiel explained.

"In your formula, you have hidden things that have a certain function. Your formula behaves in some way. When you're going to 'massage' this formula, add something or remove something . . . during your development, all of a sudden, there is a note in that formula—that sticks out, that you haven't seen before. In the formula, there's wood, vanilla, and citrus. It was all fine and nice, then you put something else in, and all of a sudden you see only the vanilla with the citrus. Or now you see the vanilla, the citrus, and the lilies, and it smells disgusting. Although it was there before, it was not the prominent thing in the smell. Even if I tell you it's in there, you don't see it. But then, I change something in the formula, and that thing sticks out. It might be an ingredient, or it might be an effect. When it's an effect, it's a nightmare because you don't know where it's from! What is it? Is that coffee, lemon, vanilla? You don't know because you've never had this scent together, and so you never know what it smells like alone."

Perfumers blend materials. They target combinations of ingredients by adding, removing, balancing. Experience tells them to know or anticipate which combination of chemicals might evoke a perceptual note different from the odor of the individual components. Conversely, this acquired knowledge also fosters a form of pattern recognition that facilitates a grasp of what elements a complex fragrance contains. Returning to the previous analysis of Cauble's wine tasting: skilled pattern recognition is what unites the perceptual expertise of perfumers and sommeliers.

The Cognitive Structure of Odor Experience

In some senses, such as vision, perceptual categorization is acquired early on in childhood, and more easily. Such perceptual categories seem intuitive and direct. Other sensory signals, especially olfactory signals, are less straightforward in their categorization. This has to do with the highly distributed and ambiguous coding of signals (Chapters 5–8). What olfactory signals mean depends on the practice of associating them with existing or new categories via additional cues (for example, cross-modal, semantic).

These associations are fundamentally learned. Learning propels the development of what Hermann von Helmholtz, in his cognitive theory of perception, had called "unconscious inferences"— inferences by which perceptions are formed through subconscious processing.[23] Unconscious inferences form against the background of prior experience and memory, immediately joined by present expectations and further shaping subsequent processes. In a sense, unconscious inferences turn into conditioned responses of our sensory system. (Helmholtz's idea is the predecessor of the "predictive brain" in current cognitive neuroscience; Chapter 8.)

Perceptions, in this context, involve judgment; they are not direct sensations. They constitute habitual response patterns inseparable from their experiential background, past and present. The workings of those unconscious inferences cannot be perceived themselves. Our mind is unable to separate its sensory content from the history of its own experience via introspection. How do people end up with relatively stable and commensurable perceptions? There are two factors at work here:

1. Our sensory systems and bodies are tuned to similar features; they share common physiological conditions while engaging in somewhat comparable enactive niches or environments.

2. We do not live in isolation: we are social creatures; we learn to engage with and communicate about things as parts of community structures on several levels (families, interest groups, societies, etcetera). Such acquired conditioning of unconscious inferences need not be accessible via introspection.

The inseparability of perception from its history of learning does not prevent it from being altered. After all, expertise is a product of targeted effort and attention. But the cognitive conditioning of perceptual experts, such as perfumers and winemakers, differs significantly from perceptual learning in naïve smellers or novices. The cognitive scientist Jon Willits, who specializes in semantic development and knowledge acquisition, outlined the difference: Experts learn through supervised training with respect to task-specific aims and conditions of success. Their categorizations are acquired concepts from other experts, and the cognitive structure is instructive to the expert's perceptual performance. Novices learn in an unsupervised, task-independent manner; their categories are socially mediated, and part of general input, not targeted presentation; and the cognitive framework is incidental to perceptual performance.[24]

The cognitive architecture of experts, therefore, varies as a consequence of differences in cognitive strategies and learning. Olfactory training in experts, unlike in novices, is not random but organized. The mapping of conceptual representations and perceptual categories onto materials is practiced. Classifiers and criteria of comparison are selective to the task at hand. And, as we saw earlier, they are also relative to expertise and material engagement, as specialists diverge in their emphasis on some dimensions over others as a result.

Diversity in the semantic space of expert perception arises from cluster learning.[25] Neither olfactory experts nor naïve smellers acquire their perceptual catalog and skills in isolation. The considerable variation in untrained performance is due to the highly unstructured acquisition of categories in odor perception. We learn about smells randomly through our behavioral landscape, often in loose association with its objects (foods, city "smellscapes," gardens, etcetera). We hardly ever learn about smells as smells. Experts, by contrast, undergo specific training of associating smells with targeted classifiers; they vary less in performance, as long as they are part of the same domain. Of course, individual variations among specialists are possible; think of specific choices of verbal descriptors in wine tasting. Such variations mirror individual differences in training and communal language conventions, while the general cognitive structuring and profiling of aroma perception via language are comparable. This is true to such a degree that individual master sommeliers will independently end up with the correct identification of the same wine.

The reason for the broader stability in domain-specific olfactory performance is that expert learning involves a model of structured clustering. Cluster learning sets out by establishing and linking classifiers and elements (a set of basic categories with benchmark odorants), and proceeds by systematic arrangements of different clusters with each other. The general process of learning is inherently relational; new materials and notions will be arranged in connection to previously acquired groups, gradually developing the semantic space of the individual. Expert learning differs inasmuch as it requires supervised, task-specific procedures and conventions. So the central processes are similar, but the tools vary, and the latter has consequences.

The perceptual structure of expert noses varies significantly from naïve smellers since they are engaging in different processes of perceptual judgment. This comes into play in two ways. First, there is the sophistication of perceptual space and its conceptual clustering, with interactive cross-cluster connections across materials and qualitative nuances. As an example, a perfumer might have memorized a repertoire of odorants with an orangey smell, such as octanal (aldehyde C-8) and octyl acetate ($C_{10}H_{20}O_2$). That perfumer can differentiate the scent of odorants, like octyl acetate, by more than one general, popular descriptor ("fruity" or "orangey") and analyze its qualitative character in much greater detail (adding nuances such as "slightly waxy"). Conversely, his or her perceptual cluster for "slightly waxy" will, next to octyl acetate, encompass several other odorants, such as allyl amyl glycolate (these instances may not be orangey but, in this case, peary and green). That perfumer further knows how to use odor clusters in chemical applications (for example, determining whether an odorant is stable enough for detergent or shampoo). In comparison with that of naïve smellers, the perceptual space of specialists is richer in detail and, notably, structured by criteria cross-connecting several layers of analysis.

Conceptual clustering in olfaction is inherently nonlinear. It radically crisscrosses material classes and perceptual categorizations. Olfactory reference, in this case, is crosscutting and ambiguous, not linear (and unlike "red denotes this range of wavelength"). Divergence in the cognitive mapping of odor (between individuals and across domains) is not a matter of subjectivity in perception. It is a result of the complexity involved in the molecular and perceptual dimensions.

Second, there is the perceptual capacity of the trained smeller, cultivated through years of observational skill and refinement. The most crucial point about olfactory expertise is that targeted training allows you to detect finer nuances in odor quality. Moreover, it enables you to extract a surplus of information as to what kinds of qualities an odorant communicates. This brings us to an observation that might not sound intuitive to armchair noses: odorants convey more than just one qualitative note.

Odorants, unlike color stimuli, frequently communicate more than one quality. Many odors yield multiple and distinct qualitative notes. Laudamiel explained: "In fragrance, ingredients always have several facets. We do not have 'the units of smells.' And I'm not sure we will ever have them, because every molecule . . . Let's put it that way. When you look at green, you see only green. You can't see yellow, no; you don't see blue. You see green. When you look at red you see only red because it activates a certain wavelength. So you're much more specific. You have one wavelength and you define the color. There is no such thing in perfumery where I smell something and that's only one thing. I don't know a single molecule. You say cut grass? Cut grass is a whole world. In cut grass is wet dirt. There is a pear note. There is a green note, which you would say is a green, leafy note. But then how do you define a green, leafy note? That's the one that smells like cut grass. So it's a catch 22. You should say that in perfumery we do not have the finished things that start somewhere and end somewhere."

Margot replied: "It's a very important thing. There are good examples; there's this one substance called ethyl citronellyl oxalate that in structural activity work is described as being musky. But now you present this blind to people, and the perfumer is going to say: 'Yeah. It's pear. It's a fruity pear. It's sweet.' Certain musk molecules also have this fruity pear component in their smell."

An odorant, in Walt Whitman's poetic words, contains multitudes. Access and appreciation of this informational richness of odorants builds on cognitive engagement through structured, iterative attention. The most vital message for theories of perception emerging from this fact is that perception is not direct, and that there is always more to things than meets the eye, or that can be grasped at first sight. Of course, we do know this from vision, where we perceive much more that we are consciously aware of (consider change blindness). Now we know this also is the case in olfaction.

We saw in previous chapters that odor perception is inherently ambiguous concerning its physical information, physiological coding, and perceptual interpretation. In response, this chapter's analysis of sensory experts now offers a renewed perspective on the psychological activity of perceptual categorization beyond the traditional approaches derived from stimulus classification.

Odors are not single, homogenous entities. A single odorant can convey various qualitative meanings. On a conceptual level, this affords your brain the opportunity to actively create odor images. They are not something reconstructed from the stimulus as mirror images, like received theories of perception suggest. This construction of conceptual content in perception is not separable from, or somehow added to, the sensory experience of an olfactory encounter. To the contrary, these processes are intricately interwoven in sensory processing.

The objective value of smell perception, accordingly, does not reside in the sameness of experience as linked to a particular stimulus. The objectivity of olfaction lies in its principles of encoding information. The causal processes that reliably connect sensory signals with adequate responses provide a much better basis to understand the notion of objectivity that characterizes olfactory perception, as well as other sensory modalities. While your nose picks up on crucial information in the external world, what that information entails, including what actions and conceptualizations it affords, profoundly depends on the evaluative dimensions, principles, and connections embodied by the coding system—in short, how the sensory system makes scents of stimulus information.

Experiential Tags in a Changing Landscape

Whether experienced by an expert or a novice, individual variations in odor perception should not be dismissed as simply subjective. They showcase the need to adopt a systems theoretical perspective in perceptual theorizing. Perceptual categorization is a learned process that incorporates skill building, allowing for varying degrees of expertise and resulting in several levels of sophistication. Odors are a creation of your brain, designed to respond adequately to information as flexible stimulus ratios in a multimodal environment, in tandem with the perceiver's physiology and its changing demands. Its causal basis of nonlinear, scrambled, and distributed processing affords smell with striking behavioral flexibility and contextually salient assignments of meaning. Your nose thus performs a highly sophisticated, personally tailored measure of the chemical environment.

The brain uses odors to choose, to draw distinctions and judge the nature of things or situations accordingly. The real challenge is to understand what kinds of choices are made via the nose, and what the mind does with odor as part of its intentional stance toward the world. Olfaction can play several cognitive roles in that context. Olfactory information is not computed into perspective-invariant perceptual objects as information in vision is.[26]

First and foremost, olfaction is about change. The nose measures slightest changes in the complex chemical composition of the environment, including a sufficient degree of flexibility regarding its contextual evaluation. In the course of percept formation, sensory input is filtered and structured by numerous other, associative, processes—including anticipation and expectation, experience and learning, memory and semantic associations, attention and awareness, cross-modal influences and physiological conditions like hunger or tiredness. What we end up perceiving as an odor, therefore, is highly dependent on a signal's combination with other processes that determine what perceptual judgments this signal affords. The same odorant can mean perceptually different things because its information is not unambiguous and it is not read as isolated pieces of information by the coding system. Odors act as indicators of a qualitative relationship between the observer and the materials from which these odors emanate and with which the observer engages.

What our noses really are good at is qualitative comparisons of odor, involving the recognition of the slightest qualitative changes and fine distinctions. Smells, in general, are qualitative measures of objects and situations; they serve an identifying function of things or environments without being coded as objects per se. An odor sniffed blindly remains sufficiently vague in terms of its source—not because smells are subjective feelings, but because of the coding and computational properties of the olfactory system (in response to the environmental promiscuity of the olfactory stimulus).

From this perspective, the sense of smell must be analyzed and modeled as a tool for organismal decision-making. Odor perceptions are distinct contextual markers in the comparison of objects or complex scenarios, not mental images "standing in" for some isolated sensory information or the coding of odors "as odors." Smell, in this context, commonly manifests distinct qualities of objects as part of a broader mental landscape. In everyday perception, the nose acts as a contextual measure. Similarly, the psychologist E. P. Köster has argued that "odors are usually best identified via the episodic memory of the situation in which they once occurred."[27]

We experience odors as perceptual tags in general cognitive processing. Think about it. While your conceptual memory for odors as odors might not be that skilled, your episodic memory of odors as associative tags for things or contexts is pretty exceptional (as in the proverbial smell of your grandmother's house). The brain tends to memorize odors with objects or situations, especially if these seem worth further consideration in future encounters and comparisons, to highlight similar-enough objects or mark certain circumstances (the good, the bad, and the really stinky). It is not the odorant itself that the brain is primed to recall, it is the contextual information that its memorization frequently aids.

That does not state that we cannot talk about odor objects. We certainly do. And we may still use notions such as "odor objects" or "odor images" for convenience to address distinct categories when talking about olfactory impressions on a conceptual level (like rose or pee). But we should not fall prey to a philosophical fallacy and think that these categories are equitable to visual objects in that they constitute robust representations of external objects in a similar sense. Odors are not isolated expressions of objects in physical stimulus space. Instead, they are processed as contextual measures of material qualities, as learned associations with regularities of stimulus patterns (relational cooccurrences).

That distinction is crucial. The richness in perceptual expressions of odor quality cannot be mapped equitably and uniformly onto stimulus space or classes as a direct consequence of the coding principles in olfaction.

This, now, provides a novel outlook for olfaction. For example, in Olofsson's view, "studying olfaction can help us understand the memory mechanisms in a new way. Declarative memory is a very developed field, obviously, compared to olfaction. But olfaction is having this very distinct kind of a perceptual organization that differs so much from the visual system. It would be quite productive to contrast those two very different types of sensory memory in terms of how they are maybe funneled into a similar kind of long-term memory storage format. This is a way where olfaction can, perhaps, reach beyond itself, and also help us understand more general cognitive functions."

The key to olfaction might be its associative memory function, with odors acting as representations of learned associations through reinforced experience. That is why an odor, when administered on its own, is so hard to identify. The way noses associate conceptual content with an odor, if not trained as an expert, is by its exposure in context, routinely with other sensory cues. And so in normal, everyday perception, smells act like experiential tags for the brain to use as a background and a pointer for consciousness, aided and abetted by cognition.

The mind, truly, has never smelled better.

CHAPTER 10

The Distillate

THE NOSE AS A WINDOW INTO MIND AND BRAIN

Olfaction offers a golden opportunity to arrive at a comprehensive theory of the senses and to clarify the premises by which we model the neural basis of information processing.

The sense of smell challenges key philosophical assumptions in theories of perception: (1) that individual variation in perceptual experience is incompatible with the objectivity of perception (Chapter 9), (2) that perceptual constancy is the chief function of sensory processing (Chapters 3 and 4), and (3) that neural topography is a fundamental organizing principle in the brain (Chapters 6–8). And so we gain revisionary insight into the brain's creation of mind.

Visuocentric theories often follow the idea that perception is all about the stable representation of objects—even though visual processing is strongly centered on motion detection. "How does one obtain constant perceptions in everyday life on the basis of continually changing sensations?" That is a good question. James J. Gibson expressed this idea.[1] Later, it was highlighted by David Marr.[2] But it characterizes *a* and not *the* function of perception. Philosophical and scientific debate yet centers on this issue because it poses stability as an expression of objectivity to demarcate the difference between appearance and reality in sensory perception. The stereotypic topography of the visual system appears to support this view. You could map the world onto your brain. The brain resembles a more or less passive platform where world and mind correlate. Neuroscience fills in the details of existing philosophical dualism between mind and world. *How* the brain arrives at structured representations of the world is an exciting research program of experimental science. *Whether* that is the function of the brain is seldom questioned.

Olfaction does not fit this correspondence model. Smell is a sense that faces an unpredictable stimulus, both in its behavioral context and physical localizability. The chemical environment is always in flux; tracking its statistics requires a simple solution to a complex problem. Having an overly specialized, predetermined map may even be disadvantageous in this context (Chapters 6 and 7). Instead, we saw the delicate border between incoming information and top-down effects repeatedly redrawn (Chapters 8 and 9). Olfaction is a strongly evaluative sense. Perception and judgment are entangled in the experience of smell. How we encounter and learn to associate things informs how we categorize them and form value judgments (Chapters 3 and 4). While world and mind are in correspondence, the brain is far from displaying the match of world and mind.

The brain is dynamic; it measures the world rather than mapping it. In this context, the brain weighs smell to answer how, what, and when to choose. Smells are an interpretation of ratios, an evaluation of signal combinations and magnitudes; they embody a representation of qualitative analogs. The brain as a measuring machine shows that perception is not primarily about stable object recognition and identification. Instead, perception is about the observational refinement of experience to flexibly evaluate the slightest qualitative differences in varying decision-making contexts. What instructs the informational content of odor situations is variable and contingent upon the associations formed between specific ratios and combinations of inputs and the expected value of their (possible) interactions. Any interpretation and the potential for perceptual generalizations of such sensory measures into perceptual categories is grounded in organismal needs, experience, and learning. It is action relative as well as memory based, and must be understood with respect to the interaction of the perceiving organism with its environment.

Once we look at the brain from the viewpoint of its activity and actions, meaning as a reasonably flexible decision-making organ, spatial representation becomes a byproduct, secondary to temporal, context-measuring behavior. With the evolutionary development and individual history of an organism as its calibration, the most fundamental principle defining perceptual processing is not stimulus representation but measurement.

The story of olfaction shows how to think through a system that doesn't accommodate philosophical intuitions, and to reconsider the conceptual foundations of neuroscience. It also opens up an alternative conception of perception.

Perception is a skill. It is a skill built on several levels, involving the neural processes and cognitive strategies by which an organism learns to smell. Perception, first and foremost, is about the measure of information in decision-making, not stability per se. How sensory information is processed into perceptual categories and cognitive objects, including what kind of classificatory activity a sensory system affords, is the very thing in question.

Perception is characterized by continuous activities of foraging and learning behavior. External information does not come in preformed categories or attached with labels. What we experience as perceptual categories is a result of the senses at work. It's an activity. Sensory input comes from various kinds of sources: surface reflections reaching the eye, pressure waves reaching the ears, volatile chemicals traveling up the nose, mechanical pressure applied to the skin, even the sensation produced by your own heartbeat. This smorgasbord of physical information is translated into neural signals, and then processed throughout several stages during which this information is filtered, structured into signaling chunks, and integrated into clusters, as well as further aligned with other, parallel processes in the nervous system. Accordingly, the perceptual content of smell must be analyzed not primarily as instances of "odor objects," but concerning "odor situations." Odor situations represent a perceptual measure of neural decision-making in context, where input cues are integrated in terms of their temporal and learned associations. From this perspective, sensory perception constitutes a measure of changing signal ratios in an environment informed by expectancy effects from top-down processes.

Where the Nose Meets the Eye

How much can olfaction inform theories of vision? When it comes to the principles of neural organization, not everything in vision that appears crystal clear actually is. Topography is less fundamental to vision than mainstream models imply. The primary visual cortex of sea turtles and other reptiles, for example, does not show any topographic organization of visual input; the visual cortex of sea turtles is nonretinotopic.[3] Perhaps it is time to find alternatives to our common model organisms—rodents and primates.

In 2017, the prominent neuroscientist Margaret Livingstone found that spatially discrete activation patterns in the fusiform face area of macaques are contingent upon experience.[4] Sensory learning is also fundamental to the arrangement of neural signals in visual information processing. These computational principles in the visual system involve more than a genetically predetermined Bauplan. The connectivity that underpins the nontopographic and associative neural signaling in olfaction offers a complementary model to explore computational principles beyond the map.

The particular principles by which a sensory system operates, feature selectivity and integration, provide the solid ground on which to determine and evaluate the characteristics of perception. For example, whether some perceptual categories, such as colors, are more discrete and less ambiguous than others, such as smells, depends on how information is first coded and further integrated into chunks. Theories of perception, therefore, must be based in the multiple interactions involved in the realization of perceptual effects. We have to ask what kinds of cognitive and behavioral mechanisms are afforded by the molecular and neural underpinnings, and vice versa.

Consider illusions in this context. Hermann von Helmholtz thought illusions were a result of unconscious inferences—the mechanisms by which the mind constructs perceptions from sensations (Chapter 9). This idea suits olfaction as well as vision. The cognitive scientist Mark Changizi suggested that many visual illusions could be explained by "neural lag."[5] An example is the Hering illusion, in which our brain expects a forward motion toward a vanishing point. The curved lines are a result of the perceptually simulated expectation of motion. This is one plausible interpretation of how the visual system makes sense of the world, which fits its neural basis and resonates with current theories of the predictive brain that integrate action-perception coupling.

Sensory perception is fundamentally shaped by development and experience, even in the context of a highly organized and, in its structure, strongly predetermined system like vision. If we hope to understand the mind through the brain, we ought to take the brain's structural and functional diversity more seriously.

Theories of Perception ≠ Theories of Vision

Reality is more than meets the eye. An overarching theory of perception is not reducible to theories of vision. The senses have diverging evolutionary histories. Vision came relatively late, with chemoreception having emerged earlier in the evolutionary pipeline. All senses participate in a variety of behavioral functions and serve these functions by different means. Why assume that vision is our best pick to understand perception? Why think that all senses operate by identical principles, for which vision is the optimal model?

Vision is a fantastic sensory system; it shaped the outlook of modern neuroscience. But the inherently visuocentric approach results in misguided chauvinism about "the other senses." The lens of vision has framed how we think about perception, its function, and conditions. It predetermined what we think perception *is*—a curious conclusion, given that we do not have sufficient understanding of what "the other senses" are, let alone how many they are. Sensory chauvinism routinely fuels neglect, especially of the more hidden systems. There are vital senses of which you are not consciously aware until you have a severe brain lesion. Contrary to popular opinion, we have more than five senses.

Modern sensory science estimates the existence of sometimes up to twenty-seven sensory modalities, depending on how you count them, and what the purpose of classification is. A number of senses exist that do not comply with the standard model of perception, including touch, proprioception, or interoception. These senses vary in their behavioral affordances, and differ in physiology and phenomenology.

Consider locomotion. Locomotion played a central role in the evolution of sensory systems. Senses like vision, olfaction, and proprioception allow organisms to move and respond to the environment. Locomotion is a central part of organismal behavior that determines how sensory information is used, including how neural systems were wired to process various types of environmental data. Not all senses deal with comparable regularities in stimulus processing, and not every sense is tuned to spatial navigation in the same way. How behavior shaped the neural organization of specific sensory systems, determining their perceptual processes, ought to differ.

A comparison of these three senses illustrates this issue. Take proprioception, the perception of one's own bodily position and movement. It links to our experience of space as physical beings navigating our bodies in space. It is a hidden sense that facilitates an unconscious understanding of spatiality through an inner sense of the body, even of one's own identity as a physical entity. Olfaction works differently; orthonasal smell allows for spatial behavior based on external cues yet lacks a comparable sense of spatial dimensionality itself. To achieve a sense of space through one's own body, proprioception might not process external cues; it is strongly coupled to visual perception. Smell, in turn, is bound to somatosensation to facilitate spatial navigation through sniffing behavior. The senses work complementarily: not always in unison but in an integrated manner.

In light of such sensory diversity, entertain, for a moment, the thought that vision might be the odd one out, not olfaction. The visual system is highly unusual. Roughly put, the visual system translates a three-dimensional environment into a two-dimensional retinal image to extract a specific set of features (shape, color, movement) before it reconstructs a three-dimensional mental image, adding depth. This is not how the computation of touch works. Nor is it how taste or other sensory systems work.

Vision is a sense with a highly specific neural organization that is directly related to its task of spatial navigation by extracting information from a predictable stimulus. Many senses, however, including touch and olfaction, developed to respond to a nonpredictable stimulus. Other senses involve regularity but not predictability, such as interoception, which is contingent upon homeostasis (a stable equilibrium in physiological processes like heart rate and hormonal balance). That is why theorizing about vision cannot be the foundation for an overarching theory of perception.

What's So Special about Smell

What nominates the nose as a model for the senses? Is smell really so different from vision? Olfaction and vision do share some perceptual effects. The "Parmesan-vomit" case at least reminds us of "Gestalt switches" in vision. Olfaction is not fundamentally different from vision. But this is a point of revolutionary implication. Olfaction does not have to be different from vision in all respects to initiate rethinking of the senses, including their participation in theories of mind and brain. The real question is what these similarities and dissimilarities between the senses are, and what their divergence means.

Throughout this book, you discovered a sensory system that operates by markedly different principles of neural representation in comparison with vision, audition, and somatosensation (Chapters 2, 5–8). Olfaction is highly variable already at the periphery, based on its genetics. It does not operate by additive stimulus coding or employ stereotypic, topographical stimulus representation. These characteristics stand in stark contrast with the idea of stimulus mapping, which has been at the core of sensory neuroscience in the twentieth century. Talk of mapping in theories of mind and brain had taken hold also in philosophical discussion, where representational theories dominate talk about perception.[6] Here, truthful perception is an accurate mental representation of physical properties. Such mapping talk, however, obfuscates what perception really is: associative learning, observational refinement, and context-sensitive decision-making. The story of the nose has shown us that common perceptual effects, including conceptual imagery (Chapters 3 and 9), hinge on the development of neural architecture more than stimulus topology.

Perceptual effects require understanding through the lens of their creation—to look explicitly at their coding, meaning their computational and neural principles. This approach is what Patricia Churchland once coined as the task of neurophilosophy: "To understand the mind means to understand the brain." To understand the brain, a focus on its structure and layout is not enough.

A Personal Distillation

Here ends the story of the olfactory brain while its investigation is still ongoing. The exploration of current trends in neuroscience brought to the fore a central philosophical lesson: Individual variation is not at odds with the notion of objectivity in perception; rather, it is an expression of the core mechanisms of sensory systems. Subjective effects are accessible, measurable, and comparable via the causal principles of their creation. The traditional dualism of objectivity and subjectivity in sensory perception presents itself as an artifact of older philosophical framing. It is time to change this thinking.

This is also a chance to slingshot naturalist philosophy into the twenty-first century, and break down the silos of institutionalized disciplinarity that are neatly dividing philosophical from neuroscientific inquiry. Philosophy must abandon the attitude that its ideas and debates are timeless and, in their application, independent of empirical developments. Neuroscience yields new philosophical questions and angles from which to revisit historically grown, deeply engrained intuitions about the mind and its creation, structure, and environment. We need to adapt our methods to the puzzle, not the other way around.

Neurophilosophy reaches further than neat words on paper; it's a practical outlook. And so this book ends on a personal note. A philosopher in a laboratory, day in and day out for three years, could not fail to leave a mark—on both sides. Beyond his scientific work, you may find Stuart Firestein in archives these days, reading nineteenth-century manuscripts. Meanwhile, my future has taken an empirical turn: I'm soon to start a lab, sparked by one fateful breakfast with Gordon Shepherd, who asked: "Have you thought of going experimental?"

Appendix
LIST OF INTERVIEWEES

Notes

Acknowledgments

Index

Appendix

LIST OF INTERVIEWEES

Acree, Terry | Food Chemist, Cornell University
08/10–11/2016, Ithaca; 04/18/2019, AChemS

Axel, Richard | Neuroscientist, Columbia University
03/07/2018, NYC

Bartoshuk, Linda | Psychologist, University of Florida
04/18/2018, AChemS

Buck, Linda | Neuroscientist, Fred Hutchinson Cancer Research
Center *09/21/2017, phone*

Doty, Richard | Clinical Scientist, University of Pennsylvania
04/18/2018, AChemS

Finger, Tom | Neuroscientist, University of Colorado
05/04/2018, Skype

Firestein, Stuart | Neuroscientist, Columbia University
05/23/2017, 01/26/2018, NYC

Frank, Marion | Psychologist, University of Connecticut
04/19/2018, AChemS

Frasnelli, Johannes | Clinical Scientist, Université du Québec à Trois-
Rivières *04/19/2018, AChemS*

Fremont, Harry | Perfumer, Firmenich
05/11/2018, NYC

Gerkin, Richard | Neuroinformatistician, Arizona State University
04/17/2018, AChemS

Gilbert, Avery	Sensory Scientist, Synesthetics *08/13/2016, Skype*
Gottfried, Jay	Neuroscientist, University of Pennsylvania *04/17/2018, AChemS*
Greer, Charles	Neuroscientist, Yale University *09/05/2017, New Haven*
Herz, Rachel	Psychologist, Brown University *04/17/2018, AChemS*
Hettinger, Thomas	Chemist, University of Connecticut *04/19/2018, AChemS*
Horowitz, Alexandra	Cognitive Scientist, Barnard College *03/06/2017, NYC*
Hummel, Thomas	Clinical Scientist, Technische Universität Dresden *04/28/2017, phone*
Kay, Leslie	Neuroscientist, University of Chicago *04/18/2018, AChemS*
Keller, Andreas	Neurogeneticist and Philosopher, CUNY *09/30/2016, 04/30/2017, NYC*
Laudamiel, Christophe	Perfumer, DreamAir *05/14/2018, Skype*
Mainland, Joel	Neuroscientist, Monell Chemical Senses Center *04/17/2018, AChemS*
Majid, Asifa	Cognitive Scientist, University of York *03/04/2018, Skype*
Margot, Christian	Chemist, Firmenich *04/17/2018, AChemS*
McGann, John	Neuroscientist, Rutgers University *04/18/2018, AChemS*
Mershin, Andreas	Biophysicist, MIT *01/27/2016, Skype*
Meyer, Pablo	IBM *04/30/2017, NYC*
Mozell, Maxwell	Physiologist, Upstate Medical University, NY, emeritus *05/03/2018, NYC*
Munger, Steve	Neuroscientist, University of Florida *04/17/2018, AChemS*

Noble, Ann C.   Sensory Chemist, University of California, Davis,
                          emeritus   *02/01/2018, phone*

Olofsson, Jonas   Cognitive Scientist, University of Stockholm
                          06/01/2017, NYC

Poivet, Erwan   Neuroscientist, Columbia University
                          09/01/2016, NYC

Puce, Aina   Neuroscientist, Indiana University Bloomington
                          03/02/2019, email

Reed, Randall   Molecular Biologist, John Hopkins University
                          04/26/2018, Skype

Rinberg, Dmitry   Neuroscientist, NYU
                          08/17/2016, NYC

Rogers, Matthew   Neurobiologist, Firmenich
                          05/18/2018, NYC

Shepherd, Gordon   Neuroscientist, Yale University
                          03/22/2017, New Haven

Smith, Barry C.   Philosopher, University of London
                          04/16–17/2017, NYC

Stopfer, Mark   Neuroscientist, NIH
                          04/18/2018, AChemS

Tauziet, Allison   Winemaker, Colgin Cellars, Napa Valley
                          01/23/2018, Skype

Vosshall, Leslie   Neuroscientist, Rockefeller University
                          02/07/2017, NYC

White, Theresa   Psychologist, Le Moyne College
                          04/27/2018, Skype

Wilson, Donald   Neuroscientist, NYU
                          01/26/2018, NYC

Wyatt, Tristram   Zoologist, University of Oxford
                          06/09/2018, Skype

Zou, Dong-Jing   Neuroscientist, Columbia University
                          04/24/2018, NYC

Notes

Preface

1. Abbé de Étienne Bonnot Condillac, *Condillac's Treatise on the Sensations*, trans. M. G. S. Carr (1754; repr., London: Favil, 1930), xxxi.
2. Immanuel Kant, *Anthropology from a Pragmatic Point of View*, ed. and trans. R. B. Louden (1798; repr., Cambridge: Cambridge University Press, 2006), 50–51.
3. Charles Darwin, *The Descent of Man, and Selection in Relation to Sex*, vol. 1 (London: Murray, 1874), 17.

Introduction

1. Alexander Graham Bell, "Discovery and Invention," Alexander Graham Bell Family Papers, 1834 to 1974: Article and Speech Files, Library of Congress, reprinted from *National Geographic Magazine*, June 1914.
2. Jutta Schickore, "Doing Science, Writing Science," *Philosophy of Science* 75, no. 3 (2008): 323–343.
3. Stuart Firestein, *Failure: Why Science Is So Successful* (New York: Oxford University Press, 2015); Stuart Firestein, *Ignorance: How It Drives Science* (New York: Oxford University Press, 2012).
4. Patricia Smith Churchland, *Neurophilosophy: Toward a Unified Science of the Mind-Brain* (Cambridge, MA: MIT Press, 1989); quote taken from

Patricia Smith Churchland, "Of Brains & Minds: An Exchange," *New York Review of Books* 61, no. 11, June 19, 2014, accessed March 20, 2019, https://www.nybooks.com/articles/2014/06/19/brains-and -minds-exchange/.

5. Paul M. Churchland, *The Engine of Reason, the Seat of the Soul: A Philosophical Journey into the Brain* (Cambridge, MA: MIT Press, 1996); John Bickle, *Psychoneural Reduction: The New Wave* (Cambridge, MA: MIT Press, 1998).

1. History of the Nose

1. Aristotle, *The Works of Aristotle*, ed. J. A. Smith and W. D. Ross, vol. 3, *On the Senses and the Sensible* (Oxford: Clarendon Press, 1931), 441b, 442b (emphasis added).

2. Theophrastus, *Enquiry into Plants and Minor Works on Odours and Weather Signs*, ed. and trans. Sir Arthur Hort (London: William Heinemann, 1916), 2:413.

3. Susan Ashbrook Harvey, *Scenting Salvation: Ancient Christianity and the Olfactory Imagination* (Berkeley: University of California Press, 2006); Christopher M. Woolgar, *The Senses in Late Medieval England* (New Haven, CT: Yale University Press, 2006).

4. Adam Hart-Davis and Emily Troscianko, *Taking the Piss: A Potted History of Pee* (Hornchurch, UK: Chalford Press, 2006), 55.

5. Sabine Krist and Wilfried Grießer, *Die Erforschung der chemischen Sinne: Geruchs- und Geschmackstheorien von der Antike bis zur Gegenwart* (Berlin: Peter Lang, 2006), 53; Robert Jütte, *A History of the Senses* (Cambridge, UK: Polity Press, 2005), 59.

6. Simon Kemp, "A Medieval Controversy about Odor," *Journal of the History of the Behavioral Sciences* 33, no. 3 (1997): 211–219; Jütte, *History of the Senses;* Woolgar, *Senses in Medieval England;* Krist and Grießer, *Erforschung der chemischen Sinne*, 55–58.

7. Kriest and Grießer, *Erforschung der chemischen Sinne*, 52; Avicenna Latinus, *Liber de anima seu sextus de naturalibus*, ed. Simone van Riet, vols. 1–3 (Leuven, Belgium: Peeters; Leiden, Netherlands: E. J. Brill, 1972), 146–153.

8. Andrea Porzionato, Veronica Macchi, and Raffaele De Caro, "The Role of Caspar Bartholin the Elder in the Evolution of the Terminology of the Cranial Nerves," *Annals of Anatomy* 195, no. 1 (2013): 28–31.

9. Woolgar, *Senses in Medieval England*, 15.

10. Carl Linnaeus and Andreas Wåhlin, *Dissertatio medica odores medicamentorum exhibens* (Stockholm: L. Salvius, 1752).

11. Carl Linnaeus, *Clavis Medicinae Duplex: The Two Keys of Medicine,* from a Swedish translation with introduction and commentary by Birger Bergh et al., trans. Peter Hogg, ed. Lars Hansen (London: Whitby, 2012).

12. Albrecht von Haller, *Elementa physiologiae corporis humani* (Lausanne: Sumptibus Marci-Michael Bousquet & Sociorum, 1757).

13. Hendrik Zwaardemaker, *Die Physiologie des Geruchs* (Leipzig: Verlag von Wilhelm Engelmann, 1895).

14. Anton Kerner von Marilaun, *The Natural History of Plants, Their Forms, Growth, Reproduction* (New York: H. Holt and Company, 1895–1896).

15. John Harvey Lovell, "Flower Odors and Their Importance to Bees: A Series of Articles," *American Bee Journal* 15 (1934): 392.

16. Frank Anthony Hampton, *The Scent of Flowers and Leaves: Its Purpose and Relation to Man* (London: Dulau, 1925).

17. G. W. Septimus Piesse, *The Art of Perfumery, and Method of Obtaining the Odors of Plants* (Philadelphia: Lindsay and Blakiston, 1857); Edward Sagarin, *The Science and Art of Perfumery* (London: McGraw-Hill, 1945); Mandy Aftel, *Essence and Alchemy: A Book of Perfume* (New York: North Point Press, 2001); Matthias Guentert, "The Flavour and Fragrance Industry—Past, Present, and Future," in *Flavours and Fragrances* (Berlin: Springer, 2007), 1–14.

18. Andrea Büttner, *Springer Handbook of Odor* (New York: Springer, 2017), 4–5.

19. Robert Boyle, *Experiments and Observations about the Mechanical Production of Odours* (London: E. Flesher, 1675).

20. Lawrence M. Principe, *The Aspiring Adept: Robert Boyle and His Alchemical Quest* (Princeton, NJ: Princeton University Press, 2000).

21. Robert Boyle, *The Philosophical Works of the Honourable Robert Boyle Esq., in Three Volumes,* ed. Peter Shaw, vol. 1 (London: W. Innys, R. Manby, and T. Longman, 1738), 412.

22. Herman Boerhaave, "Of the Smelling," in *Dr. Boerhaave's Academical Lectures on the Theory of Physic,* vol. 4 (London: W. Innys, 1745), 39–54, 40.

23. Antoine-François de Fourcroy, "Mémoire sur l'esprit recteur de Boerhaave," *Annales de chimie* 26 (1798): 232.

24. Johann Franz Simon, *Animal Chemistry with Reference to the Physiology and Pathology of Man,* vol. 2 (London: Sydenham Society, 1846), 343.

25. Friedrich Wöhler, "Ueber künstliche Bildung des Harnstoffs," *Annalen der Physik und Chemie* 88, no. 2 (1828): 253–256.

26. Günther Ohloff, Wilhelm Pickenhagen, and Philip Kraft, *Scent and Chemistry: The Molecular World of Odors* (Zürich: Wiley-VCH, 2011), 5.

27. Jean-Baptiste Dumas, "Über die vegetabilischen Substanzen, welche sich dem Campher nähert und Über einige Ätherische Öle," *Justus Liebigs Annalen der Chemie* 6, no. 3 (1833): 245–258.

28. Ohloff, Pickenhagen, and Kraft, *Scent and Chemistry*, 5.

29. Ferdinand Tiemann and Wilhelm Haarmann, "Über das Coniferin und seine Umwandlung in das aromatische Princip der Vanille," *Berichte der Deutschen Chemischen Gesellschaft* 7, no. 1 (1874): 608–623.

30. Karl Reimer, "Über eine neue Bildungsweise aromatischer Aldehyde," *Berichte der Deutschen Chemischen Gesellschaft* 9, no. 1 (1876): 423–424.

31. Firmenich, *Firmenich & Co., Successors to Chuit Naef & Co., Geneva, 1895–1945* (Geneva: Firmenich, 1945); Percy Kemp, ed., *An Odyssey of Flavors and Fragrances: Givaudan* (New York: Abrams, 2016).

32. Christopher Kemp, *Floating Gold: A Natural (and Unnatural) History of Ambergris* (Chicago: University of Chicago Press, 2012).

33. Leopold Ružička, "Die Grundlagen der Geruchschemie," *Chemiker-Zeitung* 44, no. 1 (1920): 93, 129.

34. Daniel Speich, "Leopold Ruzicka und das Verhältnis von Wissenschaft und Wirtschaft in der Chemie," *ETH History* (blog), Eidgenossische Technische Hochschule Zürich [Swiss Federal Institute of Technology, Zurich], accessed March 18, 2019, http://www.ethistory.ethz.ch /besichtigungen/touren/vitrinen/konjunkturkurven/vitrine61/.

35. Olivier Walusinski, "Joseph Hippolyte Cloquet (1787–1840)—Physiology of Smell: Portrait of a Pioneer," *Clinical and Translational Neuroscience* 2, no. 1 (2018): 2514183X17738406.

36. Hippolyte Cloquet, *Osphrésiologie, ou traité des odeurs, de sens et des organes de l'olfaction* (Paris: Chez Méquignon-Marvis, 1821).

37. Walusinski, "Joseph Hippolyte Cloquet," 6.

38. Eduard Paulsen, "Experimentelle Untersuchungen über die Strömung der Luft in der Nasenhöhle," *Sitzungsbericht der Kaiserlichen Akademie der Wissenschaften* 85 (1882): 348–373.

39. Zwaardemaker, *Physiologie des Geruchs*, 49–52.

40. Annick Le Guérer, "Olfaction and Cognition: A Philosophical and Psychoanalytic View," in *Olfaction, Taste, and Cognition*, ed. Catherine Rouby et al. (Cambridge: Cambridge University Press, 2002), 3–15.

41. Thomas Laycock, *A Treatise on the Nervous Diseases of Women* (London: Longman, Orme, Brown, Green, and Longmans, 1840).

42. Havelock Ellis, "Sexual Selection in Man: Touch, Smell, Hearing, and Vision," part 1 in *Studies in the Psychology of Sex*, vol. 2 (Philadelphia: F. A. Davis, 1905), 47–83; Constance Classen, David Howes, and Anthony Synnott, *Aroma: The Cultural History of Smell* (London: Routledge, 2002).

43. Ellis, "Sexual Selection in Man."

44. Carl Maria Giessler, *Wegweiser zu einer Psychologie des Geruches* (Hamburg, Leipzig: Leopold Voss, 1894).

45. Joel Michell, *Measurement in Psychology: A Critical History of a Methodological Concept* (Cambridge: Cambridge University Press, 1999).

46. Eleanor Acheson McCulloch Gamble, "The Applicability of Weber's Law to Smell," *American Journal of Psychology* 10, no. 1 (1898): 93.

47. Hans Henning, *Der Geruch* (Leipzig: Verlag von Johann Ambrosius Barth, 1916).

48. Eleanor Acheson McCulloch Gamble, "Taste and Smell," *Psychological Bulletin* 13, no. 3 (1916): 137. See also Eleanor Acheson McCulloch Gamble, "Review of 'Der Geruch' by Hans Henning," *American Journal of Psychology* 32, no. 2 (1921): 290–295.

49. E. C. Crocker and L. F. Henderson, "Analysis and Classification of Odors," *American Perfumer Essential Oil Review* 22 (1927): 325–327.

50. Ralf D. Bienfang, "Dimensional Characterisation of Odours," *Chronica botanica* 6 (1941): 249–250.

51. F. Nowell Jones and Margaret Hubbard Jones, "Modern Theories of Olfaction: A Critical Review," *Journal of Psychology: Interdisciplinary and Applied* 36, no. 1 (1953): 207–241.

52. Ellis, "Sexual Selection in Man." Ellis mentions the theories of von Walther (1807–1808), Zwaardemaker (1898), Haycraft (1887–1888), Rutherford (1892), Southerden (1903), and Vaschide and Van Melle (1899).

53. Malcolm Dyson, "Some Aspects of the Vibration Theory of Odour," *Perfumery and Essential Oil Record* 19 (1928): 456–459; Malcolm Dyson, "The Scientific Basis of Odour," *Journal of the Society of Chemical Industry* 57, no. 28 (1938): 647–651.

54. Robert H. Wright, "Odor and Molecular Vibration: The Far Infrared Spectra of Some Perfume Chemicals," *Annals of the New York Academy of Sciences* 116 (1964): 552–558; Robert H. Wright, "Odor and Molecular Vibration: Neural Coding of Olfactory Information," *Journal of Theoretical Biology* 64, no. 3 (1977): 473–474.

55. H. Teudt, "Eine Erklärung der Geruchserscheinungen," *Biologisches Zentralblatt* 33 (1913): 716–724.

56. M. N. Banerji, "Incidence of Smell: Theory of Surface Friction," *Indian Journal of Psychological Medicine* 6 (1930): 87–94.

57. Luca Turin, *The Secret of Scent* (London: Faber & Faber, 2006); Ann-Sophie Barwich, "How to Be Rational about Empirical Success in Ongoing Science: The Case of the Quantum Nose and Its Critics," *Studies in History and Philosophy of Science* 69 (2018): 40–51.

58. Lloyd H. Beck and Walter R. Miles, "Some Theoretical and Experimental Relationships between Infrared Absorption and Olfaction," *Science* 106 (1947): 511.

59. A. Müller, "A Dipolar Theory of the Sense of Odour," *Perfumery and Essential Oil Record* 27 (1936): 202.

60. M. Heyninx, "La physiologie de l'olfaction," *Revue d'Oto-Neuro-Ophthalmology* 11 (1933): 10–19.

61. T. H. Durrans, "The 'Residual Affinity' Odour Theory," *Perfumery and Essential Oil Record* 11 (1920): 391–393; C. E. Pressler, "Theories on Odors," *Drug and Cosmetic Industry* 62 (1948): 180–182.

62. Gertrud Woker, "The Relations between Structure and Smell in Organic Compounds," *Journal of Physical Chemistry* 10 (1906): 455–473.

63. Gösta Ehrensvärd, "Über die Primärvorgänge bei Chemozeptorenbeeinflussung," *Acta physiologica Scandinavica* 3, suppl. 9 (1942): 151.

64. J. LeMagnen, "Analyse d'odeurs complexes et homologues par fatigue," *Comptes rendus de l'Académie des Sciences* 226 (1949): 753–754; M. Ghirlanda, "Sulla presenza di glicogena nella mucosa olfattoria, puo'avere il glicogeno nasale rapporto con la funzione dell'olfatto?" *Atti dell'Accademia delle Scienze di Siena, detta dei fisiocritici* 18 (1950): 407–412.

65. J. H. Kremer, "Adsorption de matières odorantes et de narcotiques odorants par les lipoïdes," *Archives Néerlandaises de Physiologie de l'Homme et des Animaux* 1 (1916–1917): 715–725.

66. G. B. Kistiakowsky, "On the Theory of Odours," *Science* 112 (1950): 154–155.

67. John E. Amoore, "Current Status of the Steric Theory of Odor," *Annals of the New York Academy of Sciences* 116, no. 2 (1964): 457–476; John E. Amoore, *Recent Advances in Odor: Theory, Measurement, and Control* (New York: New York Academy of Sciences, 1964), 457–476; John E. Amoore, *The Molecular Basis of Odor* (Springfield, IL: Thomas, 1970).

68. Linus Pauling, "Molecular Architecture and Biological Reactions," *Chemical and Engineering News* 24, no. 10 (1946): 1375–1377.

69. Robert W. Moncrieff, "What Is Odor? A New Theory," *American Perfumer* 54 (1949): 453.

70. Günther Ohloff, *Scent and Fragrances: The Fascination of Odors and Their Chemical Perspectives,* trans. W. Pickenhagen and B. M. Lawrence (New York: Springer, 1994).

71. Günther Ohloff, "Relationship between Odor Sensation and Stereochemistry of Decalin Ring Compounds," in *Gustation and Olfaction,* ed. G. Ohloff and A. F. Thomas (Cambridge, MA: Academic Press, 1971), 178–183.

72. Ohloff, Pickenhagen, and Kraft, *Scent and Chemistry.*

73. Charles S. Sell, *Fundamentals of Fragrance Chemistry* (Weinheim, Germany: Wiley-VCH, 2019); Paolo Pelosi, *On the Scent: A Journey Through the Science of Smell* (Oxford: Oxford University Press, 2016).

74. Karen J. Rossiter, "Structure-Odor Relationships," *Chemical Reviews* 96, no. 8 (1996): 3201–3240; M. Chastrette, "Trends in Structure-Odor Relationship," *SAR and QSAR in Environmental Research* 6, no. 3–4 (1997): 215–254; Charles S. Sell, "On the Unpredictability of Odor," *Angewandte Chemie International Edition* 45, no. 38 (2006): 6254–6261.

75. Ann-Sophie Barwich, "Bending Molecules or Bending the Rules? The Application of Theoretical Models in Fragrance Chemistry," *Perspectives on Science* 23, no. 4 (2015): 443–465.

76. Maxwell M. Mozell, "The Spatiotemporal Analysis of Odorants at the Level of the Olfactory Receptor Sheet," *Journal of General Physiology* 50, no. 1 (1966): 25–41; Paul F. Kent et al., "Mucosal Activity Patterns as a Basis for Olfactory Discrimination: Comparing Behavior and Optical Recordings," *Brain Research* 981, no. 1–2 (2003): 1–11.

2. Modern Olfaction

1. Linda B. Buck and Richard Axel, "A Novel Multigene Family May Encode Odorant Receptors: A Molecular Basis for Odor Recognition," *Cell* 65, no. 1 (1991): 175–187.

2. Stuart Firestein, unpublished laudation at Harvey Society Lecture Series, May 18, 2016. Thanks to Stuart Firestein for sharing this material.

3. Stuart Firestein, Charles Greer, and Peter Mombaerts, "The Molecular Basis for Odor Recognition," *Cell* Annotated Classics, accessed

March 20, 2019, https://www.cell.com/pb/assets/raw/journals/research /cell/libraries/annotated-classics/ACBuck.pdf.

4. Stuart Firestein, "A Nobel Nose: The 2004 Nobel Prize in Physiology and Medicine," *Neuron* 45, no. 3 (2005): 333–338; Richard Axel, "Scents and Sensibility: A Molecular Logic of Olfactory Perception (Nobel Lecture)," *Angewandte Chemie International Edition* 44, no. 38 (2005): 6110–6127; Linda B. Buck, "Unraveling the Sense of Smell (Nobel Lecture)," *Angewandte Chemie International Edition* 44, no. 38 (2005): 6128–6140.

5. Ann-Sophie Barwich and Karim Bschir, "The Manipulability of What? The History of G-Protein Coupled Receptors," *Biology & Philosophy* 32, no. 6 (2017): 1317–1339; Robert J. Lefkowitz, "A Brief History of G-protein Coupled Receptors (Nobel Lecture)," *Angewandte Chemie International Edition* 52, no. 25 (2013): 6366–6378; Sara Snogerup-Linse, "Studies of G-protein Coupled Receptors. The Nobel Prize in Chemistry 2012. Award Ceremony Speech," *Royal Swedish Academy of Sciences* (2012), accessed July 9, 2017, https://www.nobelprize.org/nobel_prizes /chemistry/laureates/2012/presentation-speech.html.

6. Ann-Sophie Barwich, "What Is So Special about Smell? Olfaction as a Model System in Neurobiology," *Postgraduate Medical Journal* 92 (2015): 27–33.

7. Nicholas Wade, "Scientist at Work / Kary Mullis; After the 'Eureka,' a Nobelist Drops Out," *New York Times,* September 15, 1998, accessed December 31, 2019, https://www.nytimes.com/1998/09/15/science /scientist-at-work-kary-mullis-after-the-eureka-a-nobelist-drops-out .html; history of PCR: Paul Rabinow, *Making PCR: A Story of Biotechnology* (Chicago: University of Chicago Press, 2011).

8. Chaim Linhart and Ron Shamir, "The Degenerate Primer Design Problem," *Bioinformatics* 18, suppl. 1 (2002): S172–S181.

9. David Hubel, *Eye, Brain, and Vision,* Scientific American Library 22 (New York: W. H. Freeman, 1988); Patricia Smith Churchland, *Neurophilosophy: Toward a Unified Science of the Mind-Brain* (Cambridge, MA: MIT Press, 1989).

10. David H. Hubel and Torsten N. Wiesel, *Brain and Visual Perception: The Story of a 25-Year Collaboration* (New York: Oxford University Press, 2004).

11. Barbara Tizard, "Theories of Brain Localization from Flourens to Lashley," *Medical History* 3, no. 2 (1959): 132–145; Stanley Finger, *Origins*

of Neuroscience: A History of Explorations into Brain Function (Oxford: Oxford University Press, 2001); Erhard Oeser, *Geschichte der Hirnforschung: Von der Antike bis zur Gegenwart,* 2nd ed. (Darmstadt, Germany: WBG, 2010); S. Finger, "The Birth of Localization Theory," chap. 10 in *Handbook of Clinical Neurology* 95 (3rd series), *History of Neurology,* ed. M. Aminoff, F. Boller, and D. Swaab (Amsterdam, Netherlands: Elsevier, 2010), 117–128.

12. Lily E. Kay, *Who Wrote the Book of Life? A History of the Genetic Code* (Palo Alto, CA: Stanford University Press, 2000).

13. Stephen William Kuffler, "Neurons in the Retina: Organization, Inhibition and Excitatory Problems," *Cold Spring Harbor Symposia on Quantitative Biology* 17 (1952): 281–292; Stephen William Kuffler, "Discharge Patterns and Functional Organization of Mammalian Retina," *Journal of Neurophysiology* 16, no. 1 (1953): 37–68.

14. Jerome Y. Lettvin et al., "What the Frog's Eye Tells the Frog's Brain," *IEEE Xplore: Proceedings of the Institute of Radio Engineers* 47, no. 11 (1959): 1940–1951.

15. Gordon M. Shepherd, *Creating Modern Neuroscience: The Revolutionary 1950s* (New York: Oxford University Press, 2009).

16. Hubel, *Eye, Brain, and Vision,* 115.

17. Vernon B. Mountcastle, "Modality and Topographic Properties of Single Neurons of Cat's Somatic Sensory Cortex," *Journal of Neurophysiology* 20, no. 4 (1957): 408–434; Vernon B. Mountcastle, "Vernon B. Mountcastle," in *The History of Neuroscience in Autobiography,* vol. 6, ed. L. Squire (New York: Oxford University Press, 2009), 342–379.

18. Jennifer F. Linden and Christoph E. Schreiner, "Columnar Transformations in Auditory Cortex? A Comparison to Visual and Somatosensory Cortices?" *Cerebral Cortex* 13, no. 1 (2003): 83–89.

19. Jonathan C. Horton and Daniel L. Adams, "The Cortical Column: A Structure without a Function," *Philosophical Transactions of the Royal Society B: Biological Sciences* 360, no. 1456 (2005): 837–862.

20. Henry J. Alitto and Dan Yang, "Function of Inhibition in Visual Cortical Processing," *Current Opinion in Neurobiology* 20, no. 3 (2010): 340–346.

21. Bettina Malnic et al., "Combinatorial Receptor Codes for Odors," *Cell* 96, no. 5 (1999): 713–723; Shepherd and Firestein suggested a similar mechanism: Gordon M. Shepherd and Stuart Firestein, "Toward a Pharmacology of Odor Receptors and the Processing of Odor Images,"

Journal of Steroid Biochemistry and Molecular Biology 39, no. 4 (1991): 583–592.

22. Hans Henning, *Der Geruch* (Leipzig: Verlag von Johann Ambrosius Barth, 1916); John E. Amoore, "Specific Anosmia and the Concept of Primary Odors," *Chemical Senses* 2, no. 3 (1977): 267–281.

23. Haiqing Zhao et al., "Functional Expression of a Mammalian Odorant Receptor," *Science* 279 (1998): 237–242.

24. Michael S. Singer, "Analysis of the Molecular Basis for Octanal Interactions in the Expressed Rat I7 Olfactory Receptor," *Chemical Senses* 25, no. 2 (2000): 155–165.

25. Sandeepa Dey et al., "Assaying Surface Expression of Chemosensory Receptors in Heterologous Cells," *Journal of Visualized Experiments* 48 (2011): e2405; Hiro Matsunami, "Mammalian Odorant Receptors: Heterologous Expression and Deorphanization," *Chemical Senses* 41, no. 9 (2016): E123. For a review on the problem outline see Zita Peterlin, S. Firestein, and Matthew E. Rogers, "The State of the Art of Odorant Receptor Deorphanization: A Report from the Orphanage," *Journal of General Physiology* 143, no. 5 (2014): 527–542.

26. Joel D. Mainland et al., "Human Olfactory Receptor Responses to Odorants," *Scientific Data* 2 (2015): 150002.

27. Kerry J. Ressler, Susan L. Sullivan, and Linda B. Buck, "A Zonal Organization of Odorant Receptor Gene Expression in the Olfactory Epithelium," *Cell* 73, no. 3 (1993): 597–609.

28. Ramón y Cajal, "Studies on the Human Cerebral Cortex IV: Structure of the Olfactory Cerebral Cortex of Man and Mammals," in *Cajal on the Cerebral Cortex: An Annotated Translation of the Complete Writings*, ed. J. DeFelipe and E. G. Jones (1901 / 02; repr., New York: Oxford University Press, 1988), 289 (emphasis added).

29. Robert Vassar et al., "Topographic Organization of Sensory Projections to the Olfactory Bulb," *Cell* 79, no. 6 (1994): 981–991.

30. Gordon M. Shepherd, *Neurogastronomy: How the Brain Creates Flavor and Why It Matters* (New York: Columbia University Press, 2012), 66.

31. Edgar D. Adrian, "Olfactory Reactions in the Brain of the Hedgehog," *Journal of Physiology* 100, no. 4 (1942): 459–473; Edgar D. Adrian, "Sensory Messages and Sensation: The Response of the Olfactory Organ to Different Smells," *Acta Physiologica Scandinavica* 29, no. 1 (1953): 12–13.

32. Gordon M. Shepherd, "Gordon M. Shepherd," in *The History of Neuroscience in Autobiography*, vol. 7, ed. L. R. Squire, Society for Neuroscience,

accessed March 31, 2019, http://www.sfn.org/About/History-of
-Neuroscience/Autobiographical-Chapters.

33. Frank. R. Sharp, John S. Kauer, and Gordon M. Shepherd, "Local Sites
of Activity-Related Glucose Metabolism in Rat Olfactory Bulb during
Odor Stimulation," *Brain Research* 98, no. 3 (1975): 596–600; William B.
Stewart, John S. Kauer, and Gordon M. Shepherd, "Functional
Organization of the Rat Olfactory Bulb Analyzed by the
2-deoxyglucose Method," *Journal of Comparative Neurology* 185, no. 4
(1979): 489–495.

34. Peter Mombaerts et al., "Visualizing an Olfactory Sensory Map," *Cell*
87, no. 4 (1996): 675–686.

35. Fuqiang Xu, Charles A. Greer, and Gordon M. Shepherd, "Odor Maps
in the Olfactory Bulb," *Journal of Comparative Neurology* 422, no. 4
(2000): 489–495; Gordon M. Shepherd, W. R. Chen, and Charles A.
Greer, "Olfactory Bulb," in *The Synaptic Organization of the Brain,* ed.
G. M. Shepherd (Oxford: Oxford University Press, 2004), 165–216.

3. Minding the Nose

1. Günther Ohloff, Wilhelm Pickenhagen, and Philip Kraft, *Scent and
Chemistry: The Molecular World of Odors* (Zürich: Wiley-VCH, 2011); Paolo
Pelosi, *On the Scent: A Journey through the Science of Smell* (Oxford: Oxford
University Press, 2016).

2. Donald A. Wilson and Richard J. Stevenson, *Learning to Smell: Olfactory
Perception from Neurobiology to Behavior* (Baltimore: Johns Hopkins
University Press, 2006).

3. Gordon M. Shepherd, *Neurogastronomy: How the Brain Creates Flavor and
Why It Matters* (New York: Columbia University Press, 2012).

4. Constance Classen, David Howes, and Anthony Synnott, *Aroma: The
Cultural History of Smell* (London: Routledge, 1994); Jim Drobnick, ed.,
The Smell Culture Reader (New York: Berg, 2006).

5. Andreas Keller, *Philosophy of Olfactory Perception* (New York: Palgrave
Macmillan, 2017); Barry C. Smith, "The Nature of Sensory Experience:
The Case of Taste and Tasting," *Phenomenology and Mind* 4 (2016):
212–227; Clare Batty, "A Representational Account of Olfactory
Experience," *Canadian Journal of Philosophy* 40, no. 4 (2010): 511–538.

6. Jean-Claude Ellena, *The Diary of a Nose: A Year in the Life of a Parfumeur*
(London: Particular Books, 2012).

7. Paul H. Freedman, ed., *Food: The History of Taste* (Berkeley: University of California Press, 2007).

8. Nadia Berenstein, "Designing Flavors for Mass Consumption," *Senses and Society* 13, no. 1 (2018): 19–40.

9. Russell S. J. Keast and Andrew Costanzo, "Is Fat the Sixth Taste Primary? Evidence and Implications," *Flavour* 4, no. 1 (2015): 5.

10. Linda M. Bartoshuk, "Taste," *Stevens' Handbook of Experimental Psychology and Cognitive Neuroscience* 2 (2018): 121–154.

11. International Standards Organization, "ISO 18794:2018(en)," accessed August 17, 2019, https://www.iso.org/obp/ui/#iso:std:iso:18794:ed -1:v1:en:term:3.1.6.

12. Roger Jankowski, *The Evo-Devo Origin of the Nose, Anterior Skull Base and Midface* (Paris: Springer, 2016), chap. 6.

13. Dana M. Small et al., "Differential Neural Responses Evoked by Orthonasal versus Retronasal Odorant Perception in Humans," *Neuron* 47, no. 4 (2005): 593–605.

14. Charles Spence, "Oral Referral: On the Mislocalization of Odours to the Mouth," *Food Quality and Preference* 50 (2016): 117–128.

15. "The Truth about Youth," *McCann Worldgroup*, 2011, accessed March 24, 2019, https://mccann.com.au/wp-content/uploads/the-truth-about -youth.pdf.

16. David Melcher and Zoltán Vidnyánszky, "Subthreshold Features of Visual Objects: Unseen but Not Unbound," *Vision Research* 46, no. 12 (2006): 1863–1867.

17. Birgitta Dresp-Langley, "Why the Brain Knows More than We Do: Non-conscious Representations and Their Role in the Construction of Conscious Experience," *Brain Sciences* 2, no. 1 (2011): 1–21.

18. Andreas Keller et al., "Genetic Variation in a Human Odorant Receptor Alters Odour Perception," *Nature* 449, no. 7161 (2007): 468; Casey Trimmer et al., "Genetic Variation across the Human Olfactory Receptor Repertoire Alters Odor Perception," *Proceedings of the National Academy of Sciences* 116, no. 19 (2019): 9475–9480.

19. Nicholas Eriksson et al., "A Genetic Variant near Olfactory Receptor Genes Influences Cilantro Preference," *Flavour* 1, no. 1 (2012): 22.

20. Michael Tye, "Qualia," in *Stanford Encyclopedia of Philosophy*, ed. E. Zalta, accessed May 7, 2018, https://plato.stanford.edu/entries/qualia/.

21. Richard L. Doty, Avron Marcus, and W. William Lee, "Development of the 12-Item Cross-Cultural Smell Identification Test (CC-SIT)," *Laryngoscope* 106, no. 3 (1996): 353–356.

22. Gordon M. Shepherd, "The Human Sense of Smell: Are We Better than We Think?," *PLoS Biology* 2, no. 5 (2004): e146.

23. Yaara Yeshurun and Noam Sobel, "An Odor Is Not Worth a Thousand Words: From Multidimensional Odors to Unidimensional Odor Objects," *Annual Review of Psychology* 61 (2010): 219–241.

24. Shepherd, *Neurogastronomy,* chap. 8.

25. Tyler S. Lorig, "On the Similarity of Odor and Language Perception," *Neuroscience & Biobehavioral Reviews* 23, no. 3 (1999): 391–398.

26. Asifa Majid and Niclas Burenhult, "Odors Are Expressible in Language, as Long as You Speak the Right Language," *Cognition* 130, no. 2 (2014): 266–270.

27. Ewelina Wnuk and Asifa Majid, "Revisiting the Limits of Language: The Odor Lexicon of Maniq," *Cognition* 131, no. 1 (2014): 125–138.

28. Jonas K. Olofsson and Donald A. Wilson, "Human Olfaction: It Takes Two Villages," *Current Biology* 28, no. 3 (2018): R108–R110.

29. Andrew Dravnieks, *Atlas of Odor Character Profiles* (Philadelphia: ASTM, 1985).

30. René Magritte, *The Treachery of Images* (Los Angeles: Los Angeles County Museum of Art, 1929).

31. William G. Lycan, *Consciousness and Experience* (Cambridge, MA: MIT Press, 1996).

32. Benjamin D. Young, "Smelling Matter," *Philosophical Psychology* 29, no. 4 (2016): 520–534.

33. Ann-Sophie Barwich, "A Critique of Olfactory Objects," *Frontiers in Psychology* (June 12, 2019), http://doi.org/10.3389/fpsyg.2019.01337.

34. Tali Weiss et al., "Perceptual Convergence of Multi-component Mixtures in Olfaction Implies an Olfactory White," *Proceedings of the National Academy of Sciences* 109, no. 49 (2012): 19959–19964.

35. Karen J. Rossiter, "Structure-Odor Relationships," *Chemical Reviews* 96, no. 8 (1996): 3201–3240; Charles Sell, "On the Unpredictability of Odor," *Angewandte Chemie International Edition* 45, no. 38 (2006): 6254–6261.

36. Examples: Robert W. Moncrieff, *The Chemical Senses* (London: L. Hill, 1944); Ruth Gross-Isseroff and Doron Lancet, "Concentration-Dependent Changes of Perceived Odour Quality," *Chemical Senses* 13, no. 2 (1988): 191–204; Andrew Dravnieks, "Odor Measurement," *Environmental Letters* 3, no. 2 (1972): 81–100.

37. Henk Maarse, ed., *Volatile Compounds in Foods and Beverages* (New York and Basel: Marcel Dekker, 1991).

38. Kathleen M. Dorries et al., "Changes in Sensitivity to the Odor of Androstenone during Adolescence," *Developmental Psychobiology* 22, no. 5 (1989): 423–435.
39. Donald A. Wilson, "Pattern Separation and Completion in Olfaction," *Annals of the New York Academy of Sciences* 1170, no. 1 (2009): 306–312.
40. Dan Rokni et al., "An Olfactory Cocktail Party: Figure-Ground Segregation of Odorants in Rodents," *Nature Neuroscience* 17, no. 9 (2014): 1225.
41. Christophe Laudamiel, "The Human Sense of Smell," Center for Science and Society at Columbia University, YouTube video, accessed August 13, 2019, https://www.youtube.com/watch?v=C7uhbnRJvc8.
42. Daniel C. Dennett, *Consciousness Explained* (New York: Back Bay Books, 1991), 9.

4. How Behavior Senses Chemistry

1. Andreas A. Keller and Leslie B. Vosshall, "Human Olfactory Psychophysics," *Current Biology* 14, no. 20 (2004): R875–R878.
2. Idan Frumin et al., "A Social Chemosignaling Function for Human Handshaking," *eLife* 4 (2015): e05154.
3. Clare Batty, "A Representational Account of Olfactory Experience," *Canadian Journal of Philosophy* 40, no. 4 (2010): 511–538; Clare Batty, "What the Nose Doesn't Know: Non-veridicality and Olfactory Experience," *Journal of Consciousness Studies* 17, no. 3–4 (2010): 10–27; Clare Batty, "Smell, Philosophical Perspectives," in *Encyclopedia of the Mind*, ed. H. E. Pashler (Los Angeles: SAGE, 2013), 700–704.
4. Tim Crane, *Elements of the Mind: An Introduction to the Philosophy of Mind* (Oxford: Oxford University Press, 2001).
5. Jean-Jacques Rousseau, *Emilius and Sophia: Or, A New System of Education*, vol. 1 (London: T. Becket and P. A. de Hondt, 1763), 294.
6. Oliver Wendell Holmes, *The Autocrat of the Breakfast-Table* (Boston: Tricknor and Fields, 1865), 88; cf. Constance Classen, David Howes, and Anthony Synnott, *Aroma: The Cultural History of Smell* (London: Routledge, 1994), 87.
7. Jean-Paul Guerlain, quoted in Suzanne Biallôt, "Taking Leave of Your Senses," *Elle* 8, no. 1 (1992): 266; Jean-Paul Guerlain, quoted in Ellen Stern, "Shalimar and the House of Guerlain," *Gourmet* 56, no. 3 (1996): 84.
8. Marcel Proust, "Within a Budding Grove," in *Remembrance of Things Past*, vol. 1, *Swann's Way*, trans. C. K. Scott Moncrieff (1922; repr., New York: Modern Library, 1992), 48.

9. Crétien Van Campen, *The Proust Effect: The Senses as Doorways to Lost Memories*, trans. J. Ross (Oxford: Oxford University Press, 2014).

10. Avery N. Gilbert, *What the Nose Knows: The Science of Scent in Everyday Life* (New York: Crown Publisher, 2008), chap. 10, esp. 351.

11. Rachel S. Herz, "Trygg Engen, Pioneer of Olfactory Psychology, 1926–2009," *Chemosensory Perception* 3, no. 2 (2010): 135; Gesualdo Zucco, "Professor Trygg Engen (1926–2009)," *Chemical Senses* 35, no. 3 (2010): 181–182.

12. Harry T. Lawless and William S. Cain, "Recognition Memory for Odors," *Chemical Senses* 1, no. 3 (1975): 331–337.

13. Rachel Herz and Trygg Engen, "Odor Memory: Review and Analysis," *Psychonomic Bulletin & Review* 3, no. 3 (1996): 300–313.

14. Harry T. Lawless and Trygg Engen, "Associations to Odors: Interference, Mnemonics, and Verbal Labeling," *Journal of Experimental Psychology: Human Learning and Memory* 3, no. 1 (1977): 52.

15. Lizzie Ostrom, *Perfume: A Century of Scents* (London: Random House, 2015); Laura Eliza Enriquez, "Perfume: A Sensory Journey through Contemporary Scent," *The Senses and Society* 13, no. 1 (2018): 126–130.

16. Rachel Herz, "Perfume," in *Neurobiology of Sensation and Reward*, Frontiers in Neuroscience, ed. J. Gottfried (Boca Raton, FL: CRC Press, 2011), 371.

17. Benoist Schaal, Luc Marlier, and Robert Soussignan, "Human Foetuses Learn Odours from Their Pregnant Mother's Diet," *Chemical Senses* 25, no. 6 (2000): 729–737.

18. E.g., Paul Rozin and Edward B. Royzman, "Negativity Bias, Negativity Dominance, and Contagion," *Personality and Social Psychology Review* 5, no. 4 (2001): 296–320; Paul Rozin, Amy Wrzesniewski, and Deidre Byrnes, "The Elusiveness of Evaluative Conditioning," *Learning and Motivation* 29, no. 4 (1998): 397–415.

19. Alain Corbin, *The Foul and the Fragrant: Odor and the French Social Imagination* (Cambridge, MA: Harvard University Press, 1986).

20. Melanie A. Kiechle, *Smell Detectives: An Olfactory History of Nineteenth-Century Urban America* (Seattle: University of Washington Press, 2017).

21. K. Liddell, "Smell as a Diagnostic Marker," *Postgraduate Medical Journal* 52, no. 605 (1976): 136–138.

22. E.g., Michael McCulloch et al., "Diagnostic Accuracy of Canine Scent Detection in Early- and Late-Stage Lung and Breast Cancers," *Integrative Cancer Therapies* 5, no. 1 (2006): 30–39; Leon Frederick Campbell

et al., "Canine Olfactory Detection of Malignant Melanoma," *BMJ Case Reports* (2013): bcr2013008566.

23. Elizabeth Quigley, "Scientists Sniff Out Parkinson's Disease Smell," *BBC News,* December 18, 2017, accessed December 16, 2018, https://www.bbc.com/news/uk-scotland-42252411.

24. Jules Morgan, "Joy of Super Smeller: Sebum Clues for PD Diagnostics," *Lancet Neurology* 15, no. 2 (2016): 138–139; Drupad K. Trivedi et al., "Discovery of Volatile Biomarkers of Parkinson's Disease from Sebum," *ACS Central Science* 5 (2019): 599–606; Sarah Knapton, "Woman Who Can Smell Parkinson's Disease Helps Scientists Develop First Diagnostic Test," *Telegraph,* December 18, 2017, accessed December 16, 2018, https://www.telegraph.co.uk/science/2017/12/18/woman-can-smell-parkinsons -disease-helps- scientists-develop/?WT.mc_id=tmg_share_em.

25. Claire Guest, *Daisy's Gift: The Remarkable Cancer-Detecting Dog Who Saved My Life* (New York: Random House, 2016).

26. Ann-Sophie Barwich and Hasok Chang, "Sensory Measurements: Coordination and Standardization," *Biological Theory* 10, no. 3 (2015): 200–211.

27. Richard L. Doty, Paul Shaman, and Michael Dann, "Development of the University of Pennsylvania Smell Identification Test: A Standardized Microencapsulated Test of Olfactory Function," *Physiology & Behavior* 32, no. 3 (1984): 489–502; Daniel A. Deems et al., "Smell and Taste Disorders, a Study of 750 Patients from the University of Pennsylvania Smell and Taste Center," *Archives of Otolaryngology–Head & Neck Surgery* 117, no. 5 (1991): 519–528; Richard L. Doty, "Smell and the Degenerating Brain," *The Scientist,* October 1, 2013, accessed July 15, 2015, http://www.the-scientist .com/?articles.view/articleNo/37603/title/Smell-and-the-Degenerating -Brain/; Isabelle A. Tourbier and Richard L. Doty, "Sniff Magnitude Test: Relationship to Odor Identification, Detection, and Memory Tests in a Clinic Population," *Chemical Senses* 32, no. 6 (2007): 515–523.

28. Andreas F. Temmel et al., "Characteristics of Olfactory Disorders in Relation to Major Causes of Olfactory Loss," *Archives of Otolaryngology– Head & Neck Surgery* 128, no. 6 (2002): 635–641; Thomas Hummel, Basile N. Landis, and Karl-Bernd Hüttenbrink, "Smell and Taste Disorders," *GMS Current Topics in Otorhinolaryngology, Head and Neck Surgery* 10 (2011): Doc04.

29. Thomas Hummel et al., "'Sniffin' Sticks': Olfactory Performance Assessed by the Combined Testing of Odor Identification, Odor

Discrimination and Olfactory Threshold," *Chemical Senses* 22, no. 1 (1997): 39–52.

30. Jörn Lötsch, Heinz Reichmann, and Thomas Hummel, "Different Odor Tests Contribute Differently to the Evaluation of Olfactory Loss," *Chemical Senses* 33, no. 1 (2008): 17–21.

31. Avery N. Gilbert et al., "Olfactory Discrimination of Mouse Strains (*Mus musculus*) and Major Histocompatibility Types by Humans (*Homo sapiens*)," *Journal of Comparative Psychology* 100, no. 3 (1986): 262.

32. Claus Wedekind et al., "MHC-Dependent Mate Preferences in Humans," *Proceedings of the Royal Society B: Biological Sciences* 260, no. 1359 (1995): 245–249; Manfred Milinski, "The Major Histocompatibility Complex, Sexual Selection, and Mate Choice," *Annual Review of Ecology, Evolution, and Systematics* 37 (2006): 159–186; Andreas Ziegler, Heribert Kentenich, and Barbara Uchanska-Ziegler, "Female Choice and the MHC," *Trends in Immunology* 26, no. 9 (2005): 496–502; Claire Dandine-Roulland et al., "Genomic Evidence for MHC Disassortative Mating in Humans," *Proceedings of the Royal Society B: Biological Sciences* 286, no. 1899 (2019), https://doi.org/10.1098/rspb.2018.2664.

33. Shani Gelstein et al., "Human Tears Contain a Chemosignal," *Science* 331, no. 6014 (2011): 226–230.

34. Michael J. Russell, "Human Olfactory Communication," *Nature* 260 (1976): 520–522.

35. Original study: Martha K. McClintock, "Menstrual Synchrony and Suppression," *Nature* 229 (1971): 244–245; critical evaluation: Jeffrey C. Schank, "Menstrual-Cycle Synchrony: Problems and New Directions for Research," *Journal of Comparative Psychology* 115 (2001): 3–15; further context: Donald A. Wilson and Richard J. Stevenson, *Learning to Smell: Olfactory Perception from Neurobiology to Behavior* (Baltimore: Johns Hopkins University Press, 2006).

36. Kobi Snitz et al., "SmellSpace: An Odor-Based Social Network as a Platform for Collecting Olfactory Perceptual Data," *Chemical Senses* 44, no. 4 (2019): 267–278.

37. Shlomo Wagner et al., "A Multireceptor Genetic Approach Uncovers an Ordered Integration of VNO Sensory Inputs in the Accessory Olfactory Bulb," *Neuron* 50, no. 5 (2006): 697–709; Stephen D. Liberles and Linda B. Buck, "A Second Class of Chemosensory Receptors in the Olfactory Epithelium," *Nature* 442, no. 7103 (2006): 645.

38. Peter Karlson and Martin Lüscher, "'Pheromones': A New Term for a Class of Biologically Active Substances," *Nature* 183 (1959): 55–56.
39. Tristram D. Wyatt, "Fifty Years of Pheromones," *Nature* 457, no. 7227 (2009): 262.
40. Richard L. Doty, *The Great Pheromone Myth* (Baltimore: Johns Hopkins University Press, 2010).
41. E.g., Milos Novotny et al., "Synthetic Pheromones That Promote Inter-male Aggression in Mice," *Proceedings of the National Academy of Sciences* 82, no. 7 (1985): 2059–2061.
42. Benoist Schaal et al., "Chemical and Behavioural Characterization of the Rabbit Mammary Pheromone," *Nature* 424, no. 6944 (2003): 68.
43. Stephen D. Liberles, "Mammalian Pheromones," *Annual Review of Physiology* 76 (2014): 151–175.
44. Tristram D. Wyatt, *Pheromones and Animal Behaviour: Communication by Smell and Taste* (Cambridge: Cambridge University Press, 2003).

5. On Air

1. A conceptual comparison of "perceptual space" between various modalities is found in Ingvar Johansson, "Perceptual Spaces Are Sense-Modality-Neutral," *Open Philosophy* 1, no. 1 (2018): 14–39. I agree with Johansson that spatiality requires modality-integrative analysis but disagree on the neglect of differentiation in spatial coding across the senses.
2. John Locke, *An Essay Concerning Human Understanding* (London: Thomas Bassett, 1690).
3. Thomas Nagel, "What Is It Like to Be a Bat?" *Philosophical Review* 83, no. 4 (1974): 435–450.
4. Margaret Livingstone, *Vision and Art: The Biology of Seeing* (New York: Harry N. Abrams, 2002).
5. Mazviita Chirimuuta, *Outside Color: Perceptual Science and the Puzzle of Color in Philosophy* (Cambridge, MA: MIT Press, 2015).
6. Joel D. Mainland et al., "From Molecule to Mind: An Integrative Perspective on Odor Intensity," *Trends in Neurosciences* 37, no. 8 (2014): 443–454.
7. Yevgeniy B. Sirotin, Roman Shusterman, and Dmitry Rinberg, "Neural Coding of Perceived Odor Intensity," *eNeuro* 2, no. 6 (2015): ENEURO.0083-15.2015.

8. Daniel C. Dennett, *Consciousness Explained* (London: Penguin, 1991).

9. Neil J. Vickers et al., "Odour-Plume Dynamics Influence the Brain's Olfactory Code," *Nature* 410, no. 6827 (2001): 466; Antonio Celani, Emmanuel Villermaux, and Massimo Vergassola, "Odor Landscapes in Turbulent Environments," *Physical Review X* 4, no. 4 (2014): 041015, https://doi.org/10.1103/PhysRevX.4.041015; Ring T. Cardé and Mark A. Willis, "Navigational Strategies Used by Insects to Find Distant, Wind-Borne Sources of Odor," *Journal of Chemical Ecology* 34, no. 7 (2008): 854–866; Seth A. Budick and Michael H. Dickinson, "Free-Flight Responses of *Drosophila melanogaster* to Attractive Odors," *Journal of Experimental Biology* 209 (2006): 3001–3017.

10. Massimo Vergassola, Emmanuel Villermaux, and Boris I. Shraiman, "'Infotaxis' as a Strategy for Searching without Gradients," *Nature* 445, no. 7126 (2007): 406.

11. Noam Sobel et al., "Sniffing Longer rather than Stronger to Maintain Olfactory Detection Threshold," *Chemical Senses* 25, no. 1 (2000): 1–8; Stefan Heilmann and Thomas Hummel, "A New Method for Comparing Orthonasal and Retronasal Olfaction," *Behavioral Neuroscience* 118, no. 2 (2004): 412–419; Kai Zhao et al., "Effect of Anatomy on Human Nasal Air Flow and Odorant Transport Patterns: Implications for Olfaction," *Chemical Senses* 29, no. 5 (2004): 365–379.

12. Joel Mainland and Noam Sobel, "The Sniff Is Part of the Olfactory Percept," *Chemical Senses* 31, no. 2 (2005): 181–196.

13. M. Hasegawa and E. B. Kern, "The Human Nasal Cycle," *Mayo Clinic Proceedings* 52, no. 1 (1977): 28–34; R. Kahana-Zweig et al., "Measuring and Characterizing the Human Nasal Cycle," *PloS One* 11, no. 10 (2016): e0162918.

14. Lucia F. Jacobs et al., "Olfactory Orientation and Navigation in Humans," *PloS One* 10, no. 6 (2015): e0129387.

15. Jess Porter et al., "Mechanisms of Scent-Tracking in Humans," *Nature Neuroscience* 10, no. 1 (2007): 27.

16. Alexandra Horowitz, *Inside of a Dog: What Dogs See, Smell, and Know* (New York: Simon and Schuster, 2010).

17. Matt Wachowiak, "All in a Sniff: Olfaction as a Model for Active Sensing," *Neuron* 71, no. 6 (2011): 962–973.

18. James J. Gibson, *The Senses Considered as Perceptual Systems* (Boston: Houghton Mifflin, 1966); James J. Gibson, *The Ecological Approach to Visual Perception* (1979; repr., New York: Psychology Press, 2015).

19. Humberto R. Maturana, and Francisco J. Varela, *Autopoiesis and Cognition: The Realization of the Living* (Dordrecht, Netherlands: Springer Science & Business Media, 1991); Francisco J. Varela, Evan Thompson, and Eleanor Rosch, *The Embodied Mind: Cognitive Science and Human Experience* (Cambridge, MA: MIT Press, 2017).

20. Susan Hurley, "Perception and Action: Alternative Views," *Synthese* 129, no. 1 (2001): 3–40; Fred Keijzer, *Representation and Behavior* (Cambridge, MA: MIT Press, 2001).

21. Rufin VanRullen, "Perceptual Cycles," *Trends in Cognitive Sciences* 20, no. 10 (2016): 723–735.

22. Leslie M. Kay et al., "Olfactory Oscillations: The What, How and What For," *Trends in Neurosciences* 32, no. 4 (2009): 207–214.

23. Christina Zelano et al., "Nasal Respiration Entrains Human Limbic Oscillations and Modulates Cognitive Function," *Journal of Neuroscience* 36, no. 49 (2016): 12448–12467.

24. Rebecca Jordan et al., "Active Sampling State Dynamically Enhances Olfactory Bulb Odor Representation," *Neuron* 98, no. 6 (2018): 1214–1228.

25. Rebecca Jordan, Mihaly Kollo, and Andreas T. Schaefer, "Sniffing Fast: Paradoxical Effects on Odor Concentration Discrimination at the Levels of Olfactory Bulb Output and Behavior," *eNeuro* 5, no. 5 (2018): ENEURO.0148-18.2018.

26. Ulric Neisser, *Cognition and Reality: Principles and Implication of Cognitive Psychology* (New York: W. H. Freeman, 1976).

27. Hermann von Helmholtz, "Concerning the Perceptions in General," in *Treatise on Physiological Optics,* vol. 3, ed. and trans. J. P. C. Southall (1866; repr., New York: Optical Society of America, 1925), 1–37; Theo C. Meyering, *Historical Roots of Cognitive Science: The Rise of a Cognitive Theory of Perception from Antiquity to the Nineteenth Century* (Dordrecht, Netherlands: Springer, 2012), chaps. 7–11.

28. Paul M. Wise, Mats J. Olsson, and William S. Cain, "Quantification of Odor Quality," *Chemical Senses* 25, no. 4 (2000): 429–443; Ann-Sophie Barwich, "A Sense So Rare: Measuring Olfactory Experiences and Making a Case for a Process Perspective on Sensory Perception," *Biological Theory* 9, no. 3 (2014): 258–268.

29. Adam K. Anderson et al., "Dissociated Neural Representations of Intensity and Valence in Human Olfaction," *Nature Neuroscience* 6, no. 2 (2003): 196; Dana M. Small et al., "Dissociation of Neural Representation of Intensity and Affective Valuation in Human Gustation," *Neuron* 39,

no. 4 (2003): 701–711; Joel D. Mainland et al., "From Molecule to Mind: An Integrative Perspective on Odor Intensity," *Trends in Neurosciences* 37, no. 8 (2014): 443–454.

30. Ann-Sophie Barwich, "A Critique of Olfactory Objects," *Frontiers in Psychology* (June 12, 2019), http://doi.org/10.3389/fpsyg.2019.01337.

31. Mike W. Oram and David I. Perrett, "Modeling Visual Recognition from Neurobiological Constraints," *Neural Networks* 7, no. 6–7 (1994): 945–972.

32. Irving Biederman, "Recognition-by-Components: A Theory of Human Image Understanding," *Psychological Review* 94, no. 2 (1987): 115.

33. Yukako Yamane et al., "A Neural Code for Three-Dimensional Object Shape in Macaque Inferotemporal Cortex," *Nature Neuroscience* 11, no. 11 (2008): 1352.

34. Christopher Peacocke, "Sensational Properties: Theses to Accept and Theses to Reject," *Revue internationale de philosophie* 62, no. 242 (2008): 11.

6. Molecules to Perception

1. Alexei Koulakov et al., "In Search of the Structure of Human Olfactory Space," *Frontiers in Systems Neuroscience* 5, no. 65 (2011), http://doi.org/10.3389/fnsys.2011.00065; E. Darío Gutiérrez, Amit Dhurandhar, Andreas Keller, Pablo Meyer, and Guillermo A. Cecchi, "Predicting Natural Language Descriptions of Mono-molecular Odorants," *Nature Communications* 9, no. 1 (2018): 4979.

2. Andreas Keller et al., "Predicting Human Olfactory Perception from Chemical Features of Odor Molecules," *Science* 355, no. 6327 (2017): 820–826.

3. Andreas Keller and Leslie B. Vosshall, "Olfactory Perception of Chemically Diverse Molecules," *BMC Neuroscience* 17, no. 1 (2016): 55.

4. Ed Yong, "Scientists Stink at Reverse-Engineering Smells," *The Atlantic*, November 16, 2016, accessed July 16, 2019, https://www.theatlantic.com/science/archive/2016/11/how-to-reverse-engineer-smells/507608/.

5. Avery N. Gilbert, "Can We Predict a Molecule's Smell from Its Physical Characteristics?" *First Nerve*, February 23, 2017, accessed April 2, 2018, http://www.firstnerve.com/2017/02/can-we-predict-molecules-smell-from-its.html.

6. Kerry J. Ressler, Susan L. Sullivan, and Linda B. Buck, "A Zonal Organization of Odorant Receptor Gene Expression in the Olfactory Epithelium," *Cell* 73, no. 3 (1993): 597–609.

7. Donald A. Wilson and Richard J. Stevenson, *Learning to Smell: Olfactory Perception from Neurobiology to Behavior* (Baltimore: Johns Hopkins University Press, 2006).

8. Erwan Poivet et al., "Applying Medicinal Chemistry Strategies to Understand Odorant Discrimination," *Nature Communications* 7 (2016): 11157; Erwan Poivet et al., "Functional Odor Classification through a Medicinal Chemistry Approach," *Science Advances* 4, no. 2 (2018): eaao6086.

9. Bettina Malnic et al., "Combinatorial Receptor Codes for Odors," *Cell* 96, no. 5 (1999): 713–723.

10. Edwin A. Abbott, *Flatland: A Romance of Many Dimensions* (1884; repr., London: Penguin, 1987).

11. Lu Xu et al., "Widespread Receptor Driven Modulation in Peripheral Olfactory Coding," *bioRxiv*: 760330, accessed December 2019, https://www.biorxiv.org/content/10.1101/760330v1. This study, by Firestein's graduate student Lu Xu, was conducted during my time in the lab. The paper has yet to be published. Firestein presented the data at several occasions to the astonishment of the audience.

12. Matthew B. Bouchard et al., "Swept Confocally-Aligned Planar Excitation (SCAPE) Microscopy for High-Speed Volumetric Imaging of Behaving Organisms," *Nature Photonics* 9, no. 2 (2015): 113.

13. William S. Cain, "Odor Intensity: Mixtures and Masking," *Chemical Senses* 1, no. 3 (1975): 339–352; Douglas J. Gillan, "Taste-Taste, Odor-Odor, and Taste-Odor Mixtures—Greater Suppression within than between Modalities," *Perception & Psychophysics* 33, no. 2 (1983): 183–185; David G. Laing et al., "Quality and Intensity of Binary Odor Mixtures," *Physiology & Behavior* 33, no. 2 (1984): 309–319; Leslie M. Kay, Tanja Crk, and Jennifer Thorngate, "A Redefinition of Odor Mixture Quality," *Behavioral Neuroscience* 119, no. 2 (2005): 726–733; Larry Cashion, Andrew Livermore, and Thomas Hummel, "Odour Suppression in Binary Mixtures," *Biological Psychology* 73, no. 3 (2006): 288–297; M. A. Chaput et al., "Interactions of Odorants with Olfactory Receptors and Receptor Neurons Match the Perceptual Dynamics Observed for Woody and Fruity Odorant Mixtures," *European Journal of Neuroscience* 35, no. 4 (2012): 584–597.

14. Ricardo C. Araneda, Abhay D. Kini, and Stuart Firestein, "The Molecular Receptive Range of an Odorant Receptor," *Nature Neuroscience* 3, no. 12 (2000): 1248–1255; Yuki Oka et al., "Olfactory Receptor

Antagonism between Odorants," *EMBO Journal* 23, no. 1 (2004): 120–126; Georgina Cruz and Graeme Lowe, "Neural Coding of Binary Mixtures in a Structurally Related Odorant Pair," *Scientific Reports* 3 (2013): 1220; Peterlin Zita et al., "The Importance of Odorant Conformation to the Binding and Activation of a Representative Olfactory Receptor," *Chemistry & Biology* 15, no. 12 (2008): 1317–1327; Chaput et al., "Interactions of Odorants with Olfactory Receptors."

15. Xu et al., *Modulation in Peripheral Olfactory Coding.*

16. P. K. Stanford, *Exceeding Our Grasp: Science, History, and the Problem of Unconceived Alternatives* (Oxford: Oxford University Press, 2006).

17. Firmenich, "Firmenich Demonstrates Role of Smell to Accelerate New Toilet Economy," November 7, 2018, accessed March 29, 2019, https://www.firmenich.com/en_INT/company/news/Firmenich -demonstrates-role-of-smell-to-accelerate-new-toilet-economy.html.

18. David G. Laing and G. W. Francis, "The Capacity of Humans to Identify Odors in Mixtures," *Physiology & Behavior* 46, no. 5 (1989): 809–814; David G. Laing, "Coding of Chemosensory Stimulus Mixtures," *Annals of the New York Academy of Sciences* 510 (1987): 61–66.

19. Marion E. Frank, Dane B. Fletcher, and Thomas P. Hettinger, "Recognition of the Component Odors in Mixtures," *Chemical Senses* 42, no. 7 (2017): 537–546.

20. Thomas P. Hettinger and Marion E. Frank, "Stochastic and Temporal Models of Olfactory Perception," *Chemosensors* 6, no. 4 (2018): 44.

21. Vicente Ferreira, "Revisiting Psychophysical Work on the Quantitative and Qualitative Odour Properties of Simple Odour Mixtures: A Flavour Chemistry View. Part 2, Qualitative Aspects. A Review," *Flavour & Fragrance Journal* 27, no. 3 (2012): 201–215.

22. Madeleine M. Rochelle, Géraldine Julie Prévost, and Terry E. Acree, "Computing Odor Images," *Journal of Agricultural and Food Chemistry* 66, no. 10 (2017): 2219–2225.

7. Fingerprinting the Bulb

1. Ramón y Cajal, "Croonian Lecture: La fine structure des centres nerveux," *Proceedings of the Royal Society of London* 55 (1894): 444–468; Ramón y Cajal, "Studies on the Human Cerebral Cortex IV: Structure of the Olfactory Cerebral Cortex of Man and Mammals," in *Cajal on the Cerebral Cortex: An Annotated Translation of the Complete Writings,* ed.

J. DeFelipe and E. G. Jones (1901/02; repr., New York: Oxford University Press, 1988).

2. Steven Pinker, *How the Mind Works* (London: Penguin Books, 1998), 183.

3. John P. McGann, "Poor Human Olfaction Is a 19th-Century Myth," *Science* 356, no. 6338 (2017): eaam7263.

4. Shyam Srinivasan and Charles F. Stevens, "Scaling Principles of Distributed Circuits," *Current Biology* 29, no. 15 (2019): 2533–2540e.7.

5. Kara E. Yopak et al., "A Conserved Pattern of Brain Scaling from Sharks to Primates," *Proceedings of the National Academy of Sciences* 107, no. 29 (2010): 12946–12951.

6. Leslie M. Kay and S. Murray Sherman, "An Argument for an Olfactory Thalamus," *Trends in Neurosciences* 30, no. 2 (2007): 47–53.

7. Camillo Golgi, "Sulla fina struttura dei bulbi olfactorii," *Rivista sperimentale di freniatria e medicina legale* 1 (1875): 66–78; for an English translation of Golgi's article on the bulb see Gordon M. Shepherd et al., "The Olfactory Granule Cell: From Classical Enigma to Central Role in Olfactory Processing," *Brain Research Reviews* 55, no. 2 (2007): 373–382.

8. Gordon M. Shepherd, "Dendrodendritic Synapses: Past, Present and Future," *Annals of the New York Academy of Sciences* 1170 (2009): 215–223.

9. Mark D. Eyre, Miklos Antal, and Zoltan Nusser, "Distinct Deep Short-Axon Cell Subtypes of the Main Olfactory Bulb Provide Novel Intrabulbar and Extrabulbar GABAergic Connections," *Journal of Neuroscience* 28, no. 33 (2008): 8217–8229.

10. E.g., Zuoyi Shao et al., "Reciprocal Inhibitory Glomerular Circuits Contribute to Excitation-Inhibition Balance in the Mouse Olfactory Bulb," *eNeuro* 6, no. 3 (2019): ENEURO.0048-19.2019; Nathan N. Urban, "Lateral Inhibition in the Olfactory Bulb and in Olfaction," *Physiology & Behavior* 77, no. 4–5 (2002): 607–612; Matt Wachowiak and Michael T. Shipley, "Coding and Synaptic Processing of Sensory Information in the Glomerular Layer of the Olfactory Bulb," *Seminars in Cell & Developmental Biology* 17, no. 4 (2006): 411–423; Thomas A. Cleland, "Construction of Odor Representations by Olfactory Bulb Microcircuits," *Progress in Brain Research* 208 (2014): 177–203; Shepherd et al., "The Olfactory Granule Cell"; Alison Boyd et al., "Cortical Feedback Control of Olfactory Bulb Circuits," *Neuron* 76, no. 6 (2012): 1161–1174; Veronica Egger, Karel Svoboda, and Zachary F. Mainen, "Mechanisms of Lateral

Inhibition in the Olfactory Bulb: Efficiency and Modulation of Spike-Evoked Calcium Influx into Granule Cells," *Journal of Neuroscience* 23, no. 20 (2003): 7551–7558; Nathan N. Urban and Bert Sakmann, "Reciprocal Intraglomerular Excitation and Intra- and Interglomerular Lateral Inhibition between Mouse Olfactory Bulb Mitral Cells," *Journal of Physiology* 542, no. 2 (2002): 355–367; Christopher E. Vaaga and Gary L. Westbrook, "Distinct Temporal Filters in Mitral Cells and External Tufted Cells of the Olfactory Bulb," *Journal of Physiology* 595, no. 19 (2017): 6349–6362; Ramani Balu, R. Todd Pressler, and Ben W. Strowbridge, "Multiple Modes of Synaptic Excitation of Olfactory Bulb Granule Cells," *Journal of Neuroscience* 27, no. 21 (2007): 5621–5632.

11. Patricia Duchamp-Viret et al., "Olfactory Perception and Integration," chap. 3 in *Flavor: From Food to Behaviors, Wellbeing and Health*, ed. P. Etrévant et al., Series in Food Science, Technology and Nutrition (Cambridge, UK: Woodhead, 2016), 57–100.

12. Venkatesh N. Murthy, "Olfactory Maps in the Brain," *Annual Review of Neuroscience* 34 (2011): 233–258.

13. Thomas A. Cleland, "Early Transformations in Odor Representation," *Trends in Neuroscience* 33, no. 3 (2010): 130–139.

14. Murthy, "Olfactory Maps," 250.

15. Edward R. Soucy et al., "Precision and Diversity in an Odor Map on the Olfactory Bulb," *Nature Neuroscience* 12, no. 2 (2009): 210.

16. Soucy et al., "Precision and Diversity in an Odor Map."

17. Elissa A. Hallem and John R. Carlson, "Coding of Odors by a Receptor Repertoire," *Cell* 125, no. 1 (2006): 143–160.

18. Rainer W. Friedrich and Mark Stopfer, "Recent Dynamics in Olfactory Population Coding," *Current Opinion in Neurobiology* 11, no. 4 (2001): 468–474; Gilles Laurent, "A Systems Perspective on Early Olfactory Coding," *Science* 286, no. 5440 (1999): 723–728.

19. Paolo Lorenzon et al., "Circuit Formation and Function in the Olfactory Bulb of Mice with Reduced Spontaneous Afferent Activity," *Journal of Neuroscience* 35, no. 1 (2015): 146–160.

20. Frank R. Sharp, John S. Kauer, and Gordon M. Shepherd, "Local Sites of Activity-Related Glucose Metabolism in Rat Olfactory Bulb during Odor Stimulation," *Brain Research* 98, no. 3 (1975): 596–600; W. B. Steward et al., "Functional Organization of Rat Olfactory Bulb Analysed by the 2-Deoxyglucose Method," *Journal of Comparative Neurology* 185, no. 4 (1979): 715–734.

21. Peter Mombaerts et al., "Visualizing an Olfactory Sensory Map," *Cell* 87, no. 4 (1996): 675–686.

22. Peter Mombaerts, "Odorant Receptor Gene Choice in Olfactory Sensory Neurons: The One Receptor-One Neuron Hypothesis Revisited," *Current Opinion in Neurobiology* 14, no. 1 (2004): 31–36.

23. Kensaku Mori and Yoshihiro Yoshihara, "Molecular Recognition and Olfactory Processing in the Mammalian Olfactory System," *Progress in Neurobiology* 45, no. 6 (1995): 585–619; Naoshige Uchida et al., "Odor Maps in the Mammalian Olfactory Bulb: Domain Organization and Odorant Structural Features," *Nature Neuroscience* 3, no. 10 (2000): 1035; Kensaku Mori et al., "Maps of Odorant Molecular Features in the Mammalian Olfactory Bulb," *Physiological Reviews* 86, no. 2 (2006): 409–433.

24. Fuqiang Xu, Charles A. Greer, and Gordon M. Shepherd, "Odor Maps in the Olfactory Bulb," *Journal of Comparative Neurology* 422, no. 4 (2000): 489–495; OdorMapDB: Home-SenseLab, "Olfactory Bulb Odor Map DataBase," Yale University, accessed December 31, 2019, https://senselab.med.yale.edu/odormapdb/.

25. Leonardo Belluscio et al., "Odorant Receptors Instruct Functional Circuitry in the Mouse Olfactory Bulb," *Nature* 419, no. 6904 (2002): 296.

26. Paul Feinstein et al., "Axon Guidance of Mouse Olfactory Sensory Neurons by Odorant Receptors and the β2 Adrenergic Receptor," *Cell* 117, no. 6 (2004): 833–846.

27. Dong-Jing Zou, Alexaner Chesler, and Stuart Firestein, "How the Olfactory Bulb Got Its Glomeruli: A Just So Story?" *Nature Reviews Neuroscience* 10, no. 8 (2009): 611–618.

28. Bolek Zapiec and Peter Mombaerts, "Multiplex Assessment of the Positions of Odorant Receptor-Specific Glomeruli in the Mouse Olfactory Bulb by Serial Two-Photon Tomography," *Proceedings of the National Academy of Sciences* 112, no. 43 (2015): E5873–E5882.

29. Zapiec and Mombaerts, "Positions of Odorant Receptor-Specific Glomeruli," E5873.

30. N. Buonviso et al., "Short-Lasting Exposure to One Odour Decreases General Reactivity in the Olfactory Bulb of Adult Rats," *European Journal of Neuroscience* 10, no. 7 (1998): 2472–2475; N. Buonviso and M. Chaput, "Olfactory Experience Decreases Responsiveness of the Olfactory Bulb in the Adult Rat," *Neuroscience* 95, no. 2 (2000): 325–332.

31. Rémi Gervais et al., "What Do Electrophysiological Studies Tell Us about Processing at the Olfactory Bulb Level?" *Journal of Physiology* 101, no. 1–3 (2007): 40–45.

32. Rachel A. Ankeny and Sabina Leonelli, "What's So Special about Model Organisms?" *Studies in History and Philosophy of Science Part A* 42, no. 2 (2011): 313–323.

33. Alison Maresh et al., "Principles of Glomerular Organization in the Human Olfactory Bulb—Implications for Odor Processing," *PloS One* 3, no. 7 (2008): e2640.

34. Thomas A. Cleland and Praveen Sethupathy, "Non-topographical Contrast Enhancement in the Olfactory Bulb," *BMC Neuroscience* 7 (2006): 7.

8. Beyond Mapping, to Measuring Smells

1. Andy Clark, "Whatever Next? Predictive Brains, Situated Agents, and the Future of Cognitive Science," *Behavioral and Brain Sciences* 36, no. 3 (2013): 181–204.

2. Erich Holst and Horst Mittelstaedt, "Das Reafferenzprinzip," *Naturwissenschaften* 37, no. 20 (1950): 464–476; Roger W. Sperry, "Neural Basis of the Spontaneous Optokinetic Response Produced by Visual Inversion," *Journal of Comparative and Physiological Psychology* 43, no. 6 (1950): 482–489.

3. Ann-Sophie Barwich, "Measuring the World: Towards a Process Model of Perception," in *Everything Flows: Towards a Processual Philosophy of Biology*, ed. D. Nicholson and J. Dupré (Oxford: Oxford University Press, 2018), 227–256.

4. Christine A. Skarda and Walter J. Freeman, "How Brains Make Chaos in Order to Make Sense of the World," *Behavioral and Brain Sciences* 10, no. 2 (1987): 161–173; Walter J. Freeman, "Simulation of Chaotic EEG Patterns with a Dynamic Model of the Olfactory System," *Biological Cybernetics* 56, no. 2–3 (1987): 139–150; Yong Yao and Walter J. Freeman, "Model of Biological Pattern Recognition with Spatially Chaotic Dynamics," *Neural Networks* 3, no. 2 (1990): 153–170; Walter J. Freeman, "Neural Networks and Chaos," *Journal of Theoretical Biology* 171, no. 1 (1994): 13–18; Walter J. Freeman, "Characterization of State Transitions in Spatially Distributed, Chaotic, Nonlinear, Dynamical Systems in Cerebral Cortex," *Integrative Physiological and Behavioral Science* 29, no. 3 (1994): 294–306.

5. Anthony Chemero, "Empirical and Metaphysical Anti-representationalism," in *Understanding Representation in the Cognitive Sciences*, ed. A. Riegler, M. Peschi, and A. von Stein (Boston: Springer, 1999), 41.

6. Leslie M. Kay, Larry R. Lancaster, and Walter J. Freeman, "Reafference and Attractors in the Olfactory System during Odor Recognition," *International Journal of Neural Systems* 7, no. 4 (1996): 489–495.

7. Gilles Laurent, "Olfactory Network Dynamics and the Coding of Multidimensional Signals," *Nature Reviews Neuroscience* 3, no. 11 (2002): 884.

8. Dan D. Stettler and Richard Axel, "Representations of Odor in the Piriform Cortex," *Neuron* 63, no. 6 (2009): 854–864.

9. Dara L. Sosulski et al., "Distinct Representations of Olfactory Information in Different Cortical Centres," *Nature* 472, no. 7342 (2011): 213.

10. M. Inês Vicente and Zachary F. Mainen, "Convergence in the Piriform Cortex," *Neuron* 70, no. 1 (2011): 1–2.

11. Lewis B. Haberly, "Parallel-Distributed Processing in Olfactory Cortex: New Insights from Morphological and Physiological Analysis of Neuronal Circuitry," *Chemical Senses* 26, no. 5 (2001): 551–576; Robert L. Rennaker et al., "Spatial and Temporal Distribution of Odorant-Evoked Activity in the Piriform Cortex," *Journal of Neuroscience* 27, no. 7 (2007): 1534–1542; Donald A. Wilson and Regina M. Sullivan, "Cortical Processing of Odor Objects," *Neuron* 72, no. 4 (2011): 506–519; Donald A. Wilson, Mikiko Kadohisa, and Max L. Fletcher, "Cortical Contributions to Olfaction: Plasticity and Perception," *Seminars in Cell & Developmental Biology* 17, no. 4 (2006): 462–470; Merav Stern et al., "A Transformation from Temporal to Ensemble Coding in a Model of Piriform Cortex," *eLife* 7 (2018): e34831; Kevin A. Bolding et al., "Pattern Recovery by Recurrent Circuits in Piriform Cortex," *bioRxiv* 694331 (2019): 694331; Naoshige Uchida, Cindy Poo, and Rafi Haddad, "Coding and Transformations in the Olfactory System," *Annual Review of Neuroscience* 37 (2014): 363–385; P. Litaudon et al., "Piriform Cortex Functional Heterogeneity Revealed by Cellular Responses to Odours," *European Journal of Neuroscience* 17, no. 11 (2003): 2457–2461; Cindy Poo and Jeffry S. Isaacson, "An Early Critical Period for Long-Term Plasticity and Structural Modification of Sensory Synapses in Olfactory Cortex," *Journal of Neuroscience* 27, no. 28 (2007): 7553–7558.

12. Vicente and Mainen, "Convergence in the Piriform Cortex."

13. Chien-Fu F. Chen et al., "Nonsensory Target-Dependent Organization of Piriform Cortex," *Proceedings of the National Academy of Sciences* 111, no. 47 (2014): 16931–16936.

14. Benjamin Roland et al., "Odor Identity Coding by Distributed Ensembles of Neurons in the Mouse Olfactory Cortex," *eLife* 6 (2017): e26337.

15. John J. Hopfield, "Pattern Recognition Computation Using Action Potential Timing for Stimulus Representation," *Nature* 376, no. 6535 (1995): 33; Brice Bathellier, Olivier Gschwend, and Alan Carleton, "Temporal Coding in Olfaction," in *The Neurobiology of Olfaction*, ed. A. Menini (Boca Raton, FL: CRC Press, 2010), chapter 13.

16. Christopher D. Wilson et al., "A Primacy Code for Odor Identity," *Nature Communications* 8, no. 1 (2017): 1477.

17. Rebecca Jordan, Mihaly Kollo, and Andreas T. Schaefer, "Sniffing Fast: Paradoxical Effects on Odor Concentration Discrimination at the Levels of Olfactory Bulb Output and Behavior," *eNeuro* 5, no. 5 (2018), http://doi.org/10.1523/ENEURO.0148-18.2018.

18. Roman Shusterman et al., "Sniff Invariant Odor Coding," *eNeuro* 5, no. 6 (2018), https://doi.org/10.1523/ENEURO.0149-18.2018.

19. Gilles Laurent, Michael Wehr, and Hananel Davidowitz, "Temporal Representations of Odors in an Olfactory Network," *Journal of Neuroscience* 16, no. 12 (1996): 3837–3847; Gilles Laurent and Hananel Davidowitz, "Encoding of Olfactory Information with Oscillating Neural Assemblies," *Science* 265, no. 5180 (1994): 1872–1875; Joshua P. Martin et al., "The Neurobiology of Insect Olfaction: Sensory Processing in a Comparative Context," *Progress in Neurobiology* 95, no. 3 (2011): 427–447; Paul G. Distler and Jürgen Anthony Boeckh, "An Improved Model of the Synaptic Organization of Insect Olfactory Glomeruli," *Annals of the New York Academy of Sciences* 855, no. 1 (1998): 508–510; Elissa A. Hallem, Anupama Dahanukar, and John R. Carlson, "Insect Odor and Taste Receptors," *Annual Review of Entomology* 51 (2006): 113–135; Hugh M. Robertson, Coral G. Warr, and John R. Carlson, "Molecular Evolution of the Insect Chemoreceptor Gene Superfamily in *Drosophila melanogaster*," *Proceedings of the National Academy of Sciences* 100, no. 2 (2003): 14537–14542; Nitin Gupta and Mark Stopfer, "Insect Olfactory Coding and Memory at Multiple Timescales," *Current Opinion in Neurobiology* 21, no. 5 (2011): 768–773; Gilles Laurent et al., "Odor Encoding as an Active, Dynamical Process: Experiments, Computation, and Theory," *Annual Review of Neuroscience* 24 (2001): 263–297.

20. John G. Hildebrand and Gordon M. Shepherd, "Mechanisms of Olfactory Discrimination: Converging Evidence for Common Principles across Phyla," *Annual Review of Neuroscience* 20 (1997): 595–631; Nicholas J. Strausfeld and John G. Hildebrand, "Olfactory Systems: Common Design, Uncommon Origins?," *Current Opinion in Neurobiology* 9, no. 5 (1999): 634–639.

21. Sam Reiter and Mark Stopfer, "Spike Timing and Neural Codes for Odors," chap. 11 in *Spike Timing: Mechanisms and Function*, ed. P. M. DiLorenzo and J. D. Victor (Boca Raton, FL: CRC Press); Maxim Bazhenov and Mark Stopfer, "Forward and Back: Motifs of Inhibition in Olfactory Processing," *Neuron* 67, no. 3 (2010): 357–358.

22. Christina Zelano, Aprajita Mohanty, and Jay A. Gottfried, "Olfactory Predictive Codes and Stimulus Templates in Piriform Cortex," *Neuron* 72, no. 1 (2011): 178–187.

9. Perception as a Skill

1. Rachel S. Herz and Julia von Clef, "The Influence of Verbal Labeling on the Perception of Odors: Evidence for Olfactory Illusions?," *Perception* 30, no. 3 (2001): 381–391.

2. George Armitage Miller, "The Magical Number Seven, Plus or Minus Two: Some Limits on Our Capacity for Processing Information," *Psychological Review* 63, no. 2 (1956): 81–97.

3. Mark Sefton and Robert Simpson, "Compounds Causing Cork Taint and the Factors Affecting Their Transfer from Natural Cork Closures to Wine—A Review," *Australian Journal of Grape and Wine Research* 11, no. 2 (2005): 226–240; John Prescott et al., "Estimating a 'Consumer Rejection Threshold' for Cork Taint in White Wine," *Food Quality and Preference* 16, no. 4 (2005): 345–349.

4. Trygg Engen, *Odor Sensation and Memory* (New York: Praeger, 1991), 79.

5. William S. Cain and Bonnie Potts, "Switch and Bait: Probing the Discriminative Basis of Odor Identification via Recognition Memory," *Chemical Senses* 21, no. 1 (1996): 35–44.

6. Selig Hecht, Simon Shlaer, and Maurice Henri Pirenne, "Energy, Quanta, and Vision," *Journal of General Physiology* 25, no. 6 (1942): 819–840.

7. Denis A. Baylor, T. D. Lamb, and King-Wai Yau, "Responses of Retinal Rods to Single Photons," *Journal of Physiology* 288 (1979): 613–634.

8. Recently, debate has emerged about the nature and scope of inferences from WEIRD data; see Joseph Henrich, Steven J. Heine, and Ara Norenzayan, "Most People Are Not WEIRD," *Nature* 466, no. 7302 (2010): 29. WEIRD is an acronym for Western, educated, and from industrialized, rich, and democratic countries. In other words, theories of mind are built on data from white twenty-something college students who may or may not be representative enough from which to draw general conclusions about human nature and cognition.

9. Jean-Pierre Royet et al., "The Impact of Expertise in Olfaction," *Frontiers in Psychology* 4 (2013), https://doi.org/10.3389/fpsyg.2013.00928.

10. Johannes Frasnelli et al., "Neuroanatomical Correlates of Olfactory Performance," *Experimental Brain Research* 201, no. 1 (2010): 1–11; J. Frasnelli et al., "Brain Structure Is Changed in Congenital Anosmia," *NeuroImage* 83 (2013): 1074–1080; Janina Seubert et al., "Orbitofrontal Cortex and Olfactory Bulb Volume Predict Distinct Aspects of Olfactory Performance in Healthy Subjects," *Cerebral Cortex* 23, no. 10 (2012): 2448–2456.

11. Syrina Al Aïn et al., "Smell Training Improves Olfactory Function and Alters Brain Structure," *NeuroImage* 189 (2019): 45–54.

12. Ann-Sophie Barwich, "Up the Nose of the Beholder? Aesthetic Perception in Olfaction as a Decision-Making Process," *New Ideas in Psychology* 47 (2017): 157–165; Barry C. Smith, "Beyond Liking: The True Taste of a Wine?," *The World of Fine Wine* 58 (2017): 138–147.

13. *Somm: Into the Bottle,* directed by Jason Wise, written by Christina Tucker and Jason Wise (Los Angeles: Forgotten Man Films, 2015).

14. Ian Cauble, "somm exam," clip from *Somm: Into the Bottle,* 2015, YouTube video, 1:29, accessed March 31, 2019, https://www.youtube.com/watch?v=PKNmcCCE15E.

15. Avery N. Gilbert and Joseph A. DiVerdi, "Consumer Perceptions of Strain Differences in *Cannabis* Aroma," *PloS One* 13, no. 2 (2018): e0192247; Annette Schmelzle, "'The Beer Aroma Wheel.' Updating Beer Flavor Terminology according to Sensory Standards," *Brewing Science* 62, no. 1–2 (2009): 26–32; I. H. Suffet and P. Rosenfeld, "The Anatomy of Odour Wheels for Odours of Drinking Water, Wastewater, Compost and the Urban Environment," *Water Science and Technology* 55, no. 5 (2007): 335–344; N. P. Jolly and S. Hattingh, "A Brandy Aroma Wheel for South African Brandy," *South African Journal of Enology and Viticulture* 22, no. 1 (2001): 16–21.

16. Jancis Robinson, ed., *The Oxford Companion to Wine*, 3rd ed. (Oxford: Oxford University Press, 2006), 35–36.

17. Michael Edwards, "Fragrance Wheel," Fragrances of the World, accessed March 30, 2019, http://www.fragrancesoftheworld.com /FragranceWheel.

18. U. Harder, "Der H&R Duftkreis," *Haarman & Reimer Contact* 23 (1979): 18–27; Laura Donna, "Fragrance Perception: Is Everything Relative? Research Presents a Leap Towards a Consensus in Fragrance Mapping," *Perfumer & Flavorist* 34 (2009): 26–35.

19. David H. Pybus, "The Structure of an International Fragrance Company," chap. 5 in *The Chemistry of Fragrances: From Perfumer to Consumer*, ed. C. Sell (Cambridge: Royal Society of Chemistry, 2006).

20. Ferdinand de Saussure, *Course in General Linguistics*, ed. C. Bally et al. (1915; repr., New York: McGraw-Hill, 1966).

21. Danièle Dubois and Catherine Rouby, "Names and Categories for Odors: The Veridical Label," in *Olfaction, Taste, and Cognition*, ed. C. Rouby et al. (Cambridge: Cambridge University Press, 2002), 47–66.

22. Jean-Claude Ellena, *The Diary of a Nose: A Year in the Life of a Parfumeur* (London: Particular Books, 2012).

23. Hermann von Helmholtz, "Concerning the Perceptions in General," in *Treatise on Physiological Optics*, vol. 3, ed. and trans. J. P. C. Southall (1866; repr., New York: Optical Society of America, 1925), 1–37; Theo C. Meyering, *Historical Roots of Cognitive Science: The Rise of a Cognitive Theory of Perception from Antiquity to the Nineteenth Century* (Dordrecht, Netherlands: Springer, 2012).

24. M. N. Jones et al., "Models of Semantic Memory," in *Oxford Handbook of Mathematical and Computational Psychology*, ed. J. R. Busemeyer (Oxford: Oxford University Press, 2015), 232–254.

25. Semantic space is a specialized formatting of perceptual space, e.g., as part of a culture or expert group; distinction also in Ingvar Johansson, "Perceptual Spaces Are Sense-Modality-Neutral," *Open Philosophy* 1, no. 1 (2018): 14–39.

26. Direct comparison of odor and visual coding in Ann-Sophie Barwich, "A Critique of Olfactory Objects," *Frontiers in Psychology* (June 12, 2019), http://doi.org/10.3389/fpsyg.2019.01337.

27. Egon P. Köster, Per Møller, and Jos Mojet, "A 'Misfit' Theory of Spontaneous Conscious Odor Perception (MITSCOP): Reflections on the Role and Function of Odor Memory in Everyday Life," *Frontiers in Psychology* 5 (2014): 64.

10. The Distillate

1. James J. Gibson, *The Senses Considered as Perceptual Systems* (Boston: Houghton Mifflin, 1966).

2. David Marr, *Vision: A Computational Investigation into the Human Representation and Processing of Visual Information* (Cambridge, MA: MIT Press, 1982), 29.

3. Gilles Laurent et al., "Cortical Evolution: Introduction to the Reptilian Cortex," in *Micro-, Meso- and Macro-Dynamics of the Brain,* ed. G. Buzsáki and Y. Christen (New York: Springer, 2016), 23–33; Julien Fournier et al., "Spatial Information in a Non-retinotopic Visual Cortex," *Neuron* 97, no. 1 (2018): 164–180; R. K. Naumann et al., "The Reptilian Brain," *Current Biology* 25, no. 8 (2015): R317–R321.

4. Margaret S. Livingstone et al., "Development of the Macaque Face-Patch System," *Nature Communications* 8 (2017): 14897.

5. Mark A. Changizi et al., "Perceiving the Present and a Systematization of Illusions," *Cognitive Science* 32, no. 3 (2008): 459–503; Mark A. Changizi and David M. Widders, "Latency Correction Explains the Classical Geometrical Illusions," *Perception* 31, no. 10 (2002): 1241–1262.

6. Michael Tye, *Ten Problems of Consciousness: A Representational Theory of the Phenomenal Mind* (Cambridge, MA: MIT Press, 1997); David Pitt, "Mental Representation," in *Stanford Encyclopedia of Philosophy,* ed. E. Zalta, accessed March 30, 2019, https://plato.stanford.edu/entries/mental-representation/.

Acknowledgments

This book is the conclusion to a personal journey.

Among the many people who deserve mention, one stands out: Stuart Firestein, my mentor in neuroscience for the three years I spent in his lab at Columbia. His way of thinking altered mine. The radicality of his ideas—always questioning possible alternatives—taught me that revolutionary philosophical ideas need not come from philosophers. Still, that influence, profound as it was, cannot measure up to the deep friendship that evolved.

This friendship extends to the other members of the Firestein lab, who welcomed this oddity of a philosopher into their midst: into their lab meetings, bench work, and the annual ritual of adopting ferns. Dong-Jing Zou, Lu Xu, Erwan Poivet, Cen Zhang, Narmin Tairova, and Clara Altomare: I learned more from them than they possibly know.

Many others left their mark in the creation of this book. Meeting the mind of Terry Acree was like reading *Alice in Wonderland;* gems of ideas appeared in an unconventional format. Andreas Mershin was the first to trust me not just to think about, but also to *do* science. Gordon Shepherd planted the seeds for a philosopher willing to use her hands—to turn ideas into experiments; beyond hours of conversation on the field, he mentored me on how to pick an experimental method. Barry C. Smith opened my mind to philosophical thinking about perception through words and experience—and wine. The conversations with Chris Peacocke amplified my philosophical voice, which sometimes got hidden in scientific terminology. Andreas Keller became my comrade in revolutionary spirit. And then there is Janice Audet, my wonderful editor; she is a critical part of how this book found its shape.

How I came to meet Janice deserves a little addendum. It is representative of the unique backstory of this book and the city in which it takes place. The story began in 2015 when I traveled for the first time from New York to Germany during Christmas. In Frankfurt, I helped an elderly gentleman with his luggage on the train. He responded in English. It turned out that Peter Judd also came from New York, and we would become dear friends over the next couple of years. Our friendship included biweekly visits to the Metropolitan Opera, and during the intermission of *Fidelio* (Beethoven's only opera, and my favorite) we met friends of Peter's living in the same co-op. One of them was Joyce Seltzer, who questioned me on the way home about what we know about smell, and asked whether I'd thought of writing a book. Of course I had, and I was! Starting to, that is. She asked whether I had a proposal to show; she was an editor with Harvard University Press. The proposal needed some work, however, and Joyce walked me through my own mind one afternoon in New York, helping me to birth the heart of this book, which she then sent to her science editorial colleagues. One day later, I had heard from Janice. Like a gardener, Janice would find the parts that needed tending or cutting, that were of different varieties, or that grew out of proportion in the development of the manuscript.

At the final hour before peer review, two people read the first (and substantially longer!) manuscript, cover to cover: Aina Puce and Avery Gilbert, to whom I am deeply grateful. They often made me laugh in the long nights during the last month of finishing the manuscript. Their comments in this critical period refined my thoughts. I also thank Matthew Rodriguez and Greg Caporael for their help in the last proofing stages, Christine Hauskeller for constructive comments on the final touch of two core chapters, and Ingrid Burke for copyediting the manuscript. I am incredibly fortunate also to have had three reviewers who went beyond the ordinary with constructive commentary. One of them turned out to be John Bickle, in whom I found a passionate, kind, and exceptionally well-read soul mate in all things molecular. To the other two anonymous reviewers, evidently working in olfaction: If we ever happen to be in the same place, and you reveal yourself, please let me invite you to a drink.

A critical piece, making the trimming of the manuscript a challenge, was a trove of fascinating interviews with people willing to lend their voices to it. They are innocent of any ignorance and foolish ideas you may find. Next to

Stuart, Terry, Avery, my two Andreases, and Barry, I want to thank Charlie Greer, Linda Buck, Richard Axel, Mark Stopfer, Leslie Kay, Christian Margot, Matt Rogers, Erwan Poivet, Dong-Jing Zou, Leslie Vosshall, Theresa White, Rachel Herz, Christophe Laudamiel, Harry Fremont, Allison Tauziet, Don Wilson, Randy Reed, Dima Rinberg, Tristram Wyatt, Thomas Hummel, Steve Munger, Johannes Frasnelli, John McGann, Jay Gottfried, Joel Mainland, Rick Gerkin, Asifa Majid, Jonas Oloffson, Marion Frank, Thomas Hettinger, Linda Bartoshuk, Tom Finger, Richard Doty, Alexandra Horowitz, Ann Noble, and Pablo Meyer. Last but not least is the father of the Association for Chemoreception Sciences (AChemS): Max Mozell, a real mensch.

The work for this book was made possible by generous funding from the Presidential Scholars in Society and Neuroscience program at Columbia University. I am especially grateful to Pamela Smith. I am also genuinely thankful to Peter Todd and Jutta Schickore for picking up the tab and bringing me to Indiana University Bloomington. Here, this book was finished in a most welcoming environment. I must thank the members of the History and Philosophy of Science group, especially Jutta and Jordi Cat, who read and discussed parts of this manuscript. Last, I wish that Werner Callebaut, my friend and former mentor at the KLI Institute in Klosterneuburg, were still alive; he gave me the idea for the interviews quoted in this book.

Five people deserve mention because their discussions influenced me as a scholar beyond the book: Donna Bilak, Christine Hauskeller, Hasok Chang, Linnda Caporael, and John Dupré.

The revision of this book took place at the Duchess Anna Amalia Library in Weimar during summer 2019. This library was the perfect setting to conclude an unpredictable journey by which an undisciplined philosopher turned into a scientist.

In the intense time of writing this book, when my professional future was long uncertain, I did not have a wife and kids to thank for typing notes or doing laundry. But I do have a great family that's been letting me act out my eccentricities: my mother, my cousin Werner, and my uncle Dieter (who recently passed away and who did not tire of noting that I could have simply started with science). For better or worse, they are partly accountable for the result.

Index